GEOGRAPHIES OF MOBIL
PRACTICES, SPACES. SUB.

To Carol and Elizabeth Ann

Geographies of Mobilities:
Practices, Spaces, Subjects

Edited by

TIM CRESSWELL
University of London, UK

and

PETER MERRIMAN
Aberystwyth University, UK

Routledge
Taylor & Francis Group

LONDON AND NEW YORK

First published 2011 by Ashgate Publishing

Published 2016 by Routledge
2 Park Square, Milton Park, Abingdon, Oxon OX14 4RN
711 Third Avenue, New York, NY 10017, USA

Routledge is an imprint of the Taylor & Francis Group, an informa business

British Library Cataloguing in Publication Data
Geographies of mobilities : practices, spaces, subjects.
   1. Human locomotion. 2. Transportation. 3. Human beings –
   Migrations. 4. Human geography.
   I. Cresswell, Tim. II. Merriman, Peter.
   304.2–dc22

Library of Congress Cataloging-in-Publication Data
Cresswell, Tim.
   Geographies of mobilities : practices, spaces, subjects / edited by Tim Cresswell and Peter Merriman.
       p. cm.
   Includes bibliographical references and index.
   ISBN 978-0-7546-7316-3 (hardback)
   1. Human geography. 2. Social mobility. 3. Movement, Psychology of. 4. Migration, Internal. 5. Tourism. I. Merriman, Peter. II. Title.
   GF50.C738 2010
   304.2'3–dc22

                                                                          2010020078

ISBN: 9780754673163 (hbk)
ISBN: 9781409453659 (pbk)

Contents

List of Figures

Notes on Contributors

Peter Adey is Reader in Human Geography at Royal Holloway, University of London, UK. Inspired by the putative mobility turn his research concentrates on the processes and cultures of securitisation and militarism through the spaces and histories of the aeroplane. Two books have been recently published from these interests: *Mobility* (Routledge, 2009) and *Aerial Life: Spaces, Mobilities, Affects* (Blackwell, 2010).

John Bale is Emeritus Professor of Sports Studies at Keele University and holds honorary chairs at the University of Queensland and De Montfort University. He has written several books on geographical dimensions of sport, including *Imagined Olympians* (2002) and *Running Cultures* (2004). He is currently writing a book on *Lewis Carroll and the Idea of Sport*.

Mike Crang is Professor of Geography at Durham University. He is currently working on issues of mediated place and, mostly but not always separately, wastescapes. He is the co-editor of the journal *Tourist Studies*, and recently co-edited *Cultures of Mass Tourism: the Mediterranean in the Age of Banal Mobilities*, the *Handbook of Qualitative Geography*, the Sage *Encyclopedia of Urban Studies* and was senior editor on the *International Encyclopaedia of Human Geography*. He co-wrote *Doing Ethnographies* (with Ian Cook). He has written articles in journals in geography, literary, urban and tourism studies.

Tim Cresswell is Professor of Human Geography at Royal Holloway, University of London. He has written extensively on geographies of place and mobility and their role in the production, reproduction and transformation of society. He is currently working on an in-depth place biography of the Maxwell Street Area of Chicago's near west side and a text on geographic theory. He is the author of four books including *Place: A Short Introduction* (Blackwell, 2004) and *On the Move: Mobility in the Modern Western World* (Routledge, 2006). He has co-edited three volumes including *Gendered Mobilities* (Ashgate, 2008).

Dydia DeLyser is Associate Professor of Geography in the Department of Geography and Anthropology at Louisiana State University. Interested in qualitative research and its contemporary and historical applications to issues of landscape, memory, and mobilities, her work has appeared in journals such as the *Annals of the Association of American Geographers*, *Cultural Geographies*, *Social and Cultural Geography*, and the *Journal of Historical Geography*. Her book, *Ramona Memories: Tourism and the Shaping of Southern California*, was published by the University of Minnesota Press in 2005.

J.D. Dewsbury is Senior Lecturer in Human Geography at Bristol University. His research centres on bodies, performativity, and the concept of the event in continental philosophy, as well as on the performing arts. His recent publications have been on the work of Alain Badiou and post-phenomenological theories as he works towards a book on *Performative Spaces: Politics, Subjectivity, and the Event*.

Tim Edensor teaches Cultural Geography at Manchester Metropolitan University (UK). He is author of *Tourists at the Taj* (1998), *National Identity, Popular Culture and Everyday Life* (2002) and *Industrial Ruins: Space, Aesthetics and Materiality* (2005), the editor of *Geographies of Rhythm* (2010) and the co-editor of *Spaces of Vernacular Creativity* (2009) and *A World of Cities: Urban Theory Beyond the West* (2011). He has also written widely on tourism, urban and rural geographies, mobilities, and working-class identities. He is currently researching landscapes of illumination, geographies of rhythm and urban materiality.

Gareth Hoskins is a Lecturer in Human Geography at Aberystwyth University. He has published research on urban regeneration, memory, scale, materiality, mobility and race, as well as the role of cultural resource management protocols in the site-specific performance of heritage sites. He is currently researching the practices and spaces of competitive recollection in the sport of memorizing.

Eric Laurier is a Senior Research Fellow at the University of Edinburgh. He is currently the principal investigator on 'Assembling the line: amateur & professional work, skills and practice in digital video editing', funded by the ESRC, working together with Barry Brown and Ignaz Strebel. Although officially completed he continues to work on the ESRC project 'Habitable cars: the organisation of collective private transport' with Barry Brown and Hayden Lorimer. His interests are in practical reasoning, interaction, technology and mobility. He has written a number of papers on the uses of the car, examining how various social activities take place there, such as a parenting, and also how journeys are organised collectively by groups travelling together in cars.

Elizabeth Lee Elizabeth Lee is a post-doctoral fellow in the Department of Geography at the University of British Columbia. She is writing her dissertation on the topic of posthumous citizenship and foreign-born soldiers in the US military. Her work has been published in *Antipode* and *Environment and Planning A*. Her research interests include citizenship, military geographies, and critical race theory.

Hayden Lorimer is Senior Lecturer in Human Geography, Department of Geographical and Earth Sciences, University of Glasgow. Among other things, he does research in the open air, encompassing geographies of walking, running, rambling, collecting and field study. He has published relevant essays and articles in *Cultural Geographies*, *Social and Cultural Geography*, *Mobilities* and *Progress in Human Geography*. He walks but prefers to run.

Jo Frances Maddern is a Learning and Teaching Development Coordinator at the Centre for the Development of Staff and Academic Practice and School of Education and Lifelong Learning, Aberystwyth University. She has an interest in pedagogy in higher education, particularly theories of learning, reflective practice and academic working lives. She has a PhD in geography and also writes on issues of tourism, mobility, bio-politics, heritage, memory and identity.

Peter Merriman is Reader in Human Geography at Aberystwyth University. He has written extensively on the geography of motorways and motorway driving in Britain. His current research focuses on contemporary theorisations of space and mobility, the collaborative architecture and dance experiments of Anna and Lawrence Halprin in sixties San Francisco, and the geographies of driving in 1890s and early 1900s Britain. He is the author of *Driving Spaces: A Cultural-Historical Geography of England's M1 Motorway* (Wiley-Blackwell, 2007).

Alison Mountz is Associate Professor of Geography at Wilfrid Laurier University. She was the 2009-2010 William Lyon Mackenzie King Research Fellow with the Canada Program at Harvard University. Her latest research examines island detention centers off the shores of North America, Australia, and the European Union and is funded by a CAREER grant from the National Science Foundation. She is the author of *Seeking Asylum: Human Smuggling and Bureaucracy at the Border* (University of Minnesota Press, 2010).

David Pinder is Reader in Geography at Queen Mary, University of London. His research focuses on urbanism, city cultures and spaces of utopia. He has written about the utopianism of modernist and avant-garde movements in twentieth-century western Europe, especially the situationists; and about artistic practices, urban interventions and the politics of space. He is the author of *Visions of the City: Utopianism, Power and Politics in Twentieth Century Urbanism* (Edinburgh University Press, 2005), and a co-editor of *Cultural Geography in Practice* (Arnold, 2003).

Geraldine Pratt is Professor of Geography at the University of British Columbia. She is co-author of *Gender, Work and Space* (1995), author of *Working Feminism* (2004) and co-editor of *The Global and the Intimate* (2006) and *The Dictionary of Human Geography* (2000, 2009). She has recently co-authored a testimonial play from her research on migrant domestic workers, which was performed in Vancouver and Berlin in 2009.

Ulf Strohmayer teaches geography at the National University of Ireland, Galway, where he is currently heading the School of Geography & Archaeology. His research seeks to engage historical geographies of modernity, social theory, and urban planning. Ulf also works on questions associated with university structures in the twenty-first century. He co-authored *Key Concepts in Historical Geography* (Sage, 2009), co-edited *Geography, History and Social Science* (Kluwer, 1995), *Space and Social Theory. Interpreting Modernity and Postmodernity* (Blackwell, 1997) and *Human Geography: A Century Revisited* (Arnold 2004) and has published widely in geographical journals.

Chapter 1

Introduction:
Geographies of Mobilities – Practices, Spaces, Subjects

Tim Cresswell and Peter Merriman

What can geography offer to a 'new mobilities paradigm' (Sheller and Urry 2006)? The very question suggests that geography needs to embrace mobility. Sociology, anthropology and other disciplines across the social sciences and the humanities have gone mobile (Urry 2007; Urry 2000; Clifford 1997; Kaufmann 2002). Why not follow suit? In many ways, of course, we have. Geographers are leading contributors to and editors of the journal *Mobilities*. We have our own mobilities text book (Adey 2009) and monographs on mobility in general (Cresswell 2006) and specific forms of moving (Merriman 2007). Our conferences are jam packed with sessions with mobility or mobilities in the title.

Equally, it could reasonably be argued that we have no need to embrace a 'new mobilities paradigm' because we have always had mobility as a central focus of work in human geography. Indeed, a call for a new mobilities paradigm in our discipline has often been repeated. In 1938, for instance, the Scottish geographer Percy Crowe, in an argument for a 'progressive geography', suggested that we had become too focused on fixed things and needed to pay attention to process and circulation (Crowe 1938). Geographers, he argued, had 'advanced a static geography ... incapable of seeing movement except as pattern', but a future 'dynamic' geography must adjust its focus to study 'men and things moving' (Crowe 1938, 14):

> The study of things moving will at least take us a step along the right road, for, as compared with static distribution, movement implies three essentials – origin, destination, and an effective will to move. Movement does not take place in a vacuum, it is effected upon the surface of the earth and it is very largely through movement that Mankind becomes conscious of its geography. (Crowe 1938, 14)

Crowe's call was largely ignored, but it was repeated with the move to a 'theoretical geography' of space (rather than region) that we now know as the quantitative revolution. One proponent of such a mobile human geography was William Bunge.

He suggested that it was movement and pattern which lay at the heart of a new 'theoretical geography' that could cut across the human and physical domains:

> Notice that any explanation of a location involves the notion of movement. Even such static features as mountains and sea coasts are explained by movements over long geologic periods. In many ways patterns and movements are interrelated as are the chicken and the egg with one causing the other. Does the location of the river valley cause the movement of the river or is it the other way? Obviously one operates on the other. Thus, theoretical geography, the geography of explanation, is interested in both movements and patterns. (Bunge 1966, xvi)

In his manifesto for a 'Theoretical Geography' Bunge argued that movement was the key geographical 'fact' to be explored, modelled, theorized and explained. It was in the study of movement, he argued, that theoretical geography had made its most significant advances. Chapter 5 of his book is called 'Toward a General Theory of Movement'. It is rooted in the question posed by Edward Ullman – 'What makes objects move over the earth's surface?'

> It can be argued that Ullman's question encompasses all geographic theory, since in explaining how an object acquires its location it is difficult to avoid the notion of movement. Even such "static" features as mountains and seacoasts are explained in terms of movements taking place over long periods. (Bunge 1966, 112)

At a time when spatial scientists were looking to theoretical approaches to space and time in physics and mathematics (see, e.g. Harvey 1969), Bunge suggested that the movement of people could be equated to the movement of electricity or the flows of fluids, as understood by scientists. To a positivist such as Bunge (in *Theoretical Geography* anyway) it is important that generalizations are as general as possible. So laws that could be applied to people and to water were particularly valued. Movement, then, was incorporated into a general mathematical geography. It was positioned at the heart of the project of spatial science.

This was never clearer than in the development of transport geography. The study of transport became a central part of geography during the late 1960s and early 1970s when the urge to quantify and produce 'laws' was at its height. Gravity models and spatial interaction theory were both used to understand and predict transport-aided movement. A 'rational-mobile-person' was invented who was seen to make careful decisions about when and how to move. Here is how this is described at the end of an influential transport geography text.

> Transport exists for the purpose of bridging spatial gaps, though these gaps can be expressed not only in terms of distance but also of time and cost. It is the means by which people and goods can be moved from the place where they are at the moment to another place where they will be at a greater advantage; goods

can be sold at a higher price, people can get a better job, or live in the sort of house they prefer, or go for a holiday at the seaside. In short, people and goods are transported from one place where their utility is lower to another where it is higher. Transport as a fundamental human activity may thus be effectively studied in spatial terms: geographical methods are basic to such study and are of practical relevance to the solution of many of the problems associated with the transport industry and with its activities. (White and Senior 1983, 207)

Occasionally, however, other marginal figures would appear as quaint and unpredictable exceptions to law-like behaviour. Three such exceptions occur at the beginning of the same text book.

> For the most part the demand for transport is *derived*. With exceptions such as motorists who simply drive into the country, passengers on cruise liners and 'railfans', transport is used as a means to an end: the movement of people and goods from where they are to another place where, for the time being at any rate, their satisfaction of value will be enhanced. Transport creates *utilities of place*. (White and Senior 1983, 1)

While movement is most often seen as the outcome of rational choices involving the comparison of one location or mode with another, there are people such as 'the leisure motorist', 'cruise ship passenger' of 'trainspotter' who conduct mobility for its own sake. These people are beyond the scope of spatial science. They are scientific anomalies, or perhaps simply irrational. Needless to say, they are irrelevant to the scientific approach.

This is not the only place that the differently-mobile appear as exceptions and deviations. Abler, Adams and Gould also note the always-awkward exceptions to general rules of movement. 'Least net effort' is one of the important organizing principles of many approaches to mobility in spatial science. It suggests that there is (or perhaps, should be) a general drive to reduce the amount of effort spent in moving from place to place.

> We stress *net* effort to emphasize that the very movement process itself may carry benefits at the same time that costs are incurred. A commuter sitting in a traffic jam inhaling gasoline and carbon monoxide fumes pays a high cost for his trip, but he also has relative peace and quiet twice a day, a radio to listen to, and the feeling that for a while at least he is the boss. If his job were to move next door to his house he would probably move. (Abler et al. 1971, 253)

This is but one example of how movement is presented as a secondary geographical fact made necessary by the arrangement of primary considerations of space and location. So despite Crowe's call for geographers to focus on 'men and things moving', rather than the nodes or networks that provide the material infrastructure for such movement, geographers who worked obsessively on movement continued to relegate it as logically secondary to the arrangements of space and place. This is why the apparently throw-away references to patterns of mobility that do not fit the various models put forward by spatial scientists catch the attention. Qualitative exceptions, differences and experiences of movement leap off the page, but spatial scientists relegate such details to footnotes and asides. In contrast, *we* see the qualities of mass movements, as well as marginalized or purportedly 'irrational' movements, as important and worthy of study. *We* are interested in producing critical analyses of these practices, spaces and subjects, whether the 'motorists who simply drive into the country, passengers on cruise liners and "railfans"' (White and Senior 1983, 1), or the commuter enjoying 'relative peace and quiet twice a day, a radio to listen to, and the feeling that for a while at least he is the boss' (Abler et al. 1971, 253). Turning these experiences into a footnote is a result of thinking of movement as a cost and as dead time.

Mobilities Again

In many ways, then, geographers are not coming to mobilities anew but are revisiting an old friend. In our discussion thus far, we have concentrated on spatial science and transport geography, but there are many other ways in which geographers have positioned things or people on the move as central to the discipline. These include the development of migration theory in population and development geography (Boyle and Halfacree 1998), the tradition of time-space geographies associated with the Lund School (Hägerstrand and Pred 1981; Pred 1977), accounts of journeys to work developed within feminist geography (Hanson and Pratt 1995; Law 1999), geographies of tourism (Crouch 1999), choreographies of place developed within the phenomenological tradition (Seamon 1979), historical geographies of transport and mobility (Freeman and Aldcroft 1985; Thrift 1990; Freeman 1999; Pirie 2009), geographies of commodities, globalization and capital (Allen and Pryke 1999; Cook 2004), research on the imaginative geographies of travel writing (Blunt 1994; Duncan and Gregory 1999), and accounts of the role of travel and exploration in the very foundation of our discipline (Driver 1999; Stoddart 1986; Domosh, 1991).

Despite all of this, however, it is still the case that geographical knowledge often assumes a stable point of view, a world of places and boundaries and territories rooted in time and bounded in space. A new focus on mobilities in geography allows us to re-centre it in the discipline. While drawing on the traditions of mobility research noted above, geographies of mobility often start with the fact of moving and retain that as a focus. The apparently marginal mobilities of spatial science (the trainspotter etc.) become central to our investigations. We do not want to leave the commuter listening to the radio as a marginal curiosity. Rather we want to make her central to our interests by asking exactly what happens on the move. How is mobile time and space filled with liveliness? The mobile worlds that are labelled dead, irrational and dysfunctional by transport geographers and others come alive when they become the focus of our attention.

This book is divided into three sections that reflect the subtitle – practices, spaces and subjects. We recognize at the outset that every chapter includes aspects of all three. How could it be possible to write about the practice of driving without considering the space of the road or the subject position 'driver'? We have simply asked authors to start from a departure point and journey from there.

Practices

Mobility is practiced, and practice is often conflated with mobility. To move is to do something. Moving involves making a choice within, or despite, the constraints of society and geography. It is no surprise, therefore, that in Michel de Certeau's oft-cited classic *The Practice of Everyday Life* (de Certeau 1984), he focuses on the act of walking in the city in order to elucidate the tactical practices of the weak. Staying still (insofar as such a thing is possible) is also a notable practical positioning in the face of surrounding mobilities and the compulsion to move. An attention to the practice and performance of mobilities forms an important component of recent work on the geographies of mobilities, and the philosophical agendas driving much of the 'new mobilities paradigm' are inspired by a post-structuralist sensitivity to movement and practice. Indeed, in outlining and defining what he terms 'non-representational theory', Nigel Thrift has referred to *it* (though, elsewhere, he stresses the plurality of such theories) as a 'theory of mobile practices' (Thrift 2000, 556), and he has shown how there is an 'almost/not quite ontology which is gathering momentum around the key trope of mobility' (Thrift 1996, 258). Mobile, embodied practices are central to how we experience the world, from practices of writing and sensing, to walking and driving. Our mobilities create spaces and stories – spatial stories.

In this section the authors examine five modes of mobile practice: walking; running; dancing; driving; and flying. These practices are associated with different spaces and scales of movement. They involve a range of embodied engagements and an array of technologies and infrastructures. As the chapters by Hayden Lorimer and John Bale show, even running and walking – practices associated with the physical capacities of animate bodies – have become embroiled with a range of more-or-less complex technologies, from the shoe, running trainer and walking boot, to the asphalt running track, digital stopwatch, personal stereo, rucksack and map (see also Matless 1998; Michael 2000; Bull 2000). Driving and flying are practices which have clearly become dependent upon an extensive network of technologies and spaces, from different types and makes of airplane and motor vehicle, to the spaces of the road, motorway, car park, airport and the sky. These practices and associated spaces are entwined with a complex array of political, cultural, economic and environmental debates. All of these embodied mobile practices have complex histories and geographies, as has been exemplified by recent work on dance in geography (Nash 2000; Cresswell 2006). These practices have also come to be associated with different ways of being and thinking, and different ethics, aesthetics and ecologies. Walking has been variously constructed as romantic, reflective, escapist, natural; running as efficient, powerful, or exhausting; dancing as elegant, poetic, fun or embarrassing; driving as modern, essential, stressful, dangerous or environmentally destructive; flying as wonderous, modern, and (again) polluting. While there is a focus, in this section, on embodied practices which appear to be active and controlled (rather than passive), we should not forget that such actions as standing still, sitting or being a 'passenger' in a car, train or plane are equally, if differently, active in their embodied practices, affects, mobilities and fixities (cf. Laurier and Brown 2008; Laurier et al. 2008; Bissell 2010).

In Chapter 2 Hayden Lorimer provides a critical overview of the geographies of walking studies. He discusses how walks have been presented as a product of places, as features of everyday life, adopted as an artistic practice, and used as a way to interrogate the relationship between self and landscape, or self and world. He shows how walking has been adopted as an embodied methodological practice as well as a practice to observe or study, and how walks have been the subject of historical geographic study as well as integral to contemporary phenomenologies of landscape. Walking as a historic practice, artistic method and contemporary philosophical aid appears to connect important themes which lie at the heart of geography: embodiment, landscape, place, experience, practice, mobility, representation, materiality, subjectivity, objectivity.

In Chapter 3 John Bale addresses the practice of running. Running is frequently presented as a leisure pursuit, but in this chapter Bale shows how competitive, 'sportised' running is perhaps best considered as a form of work. Over the past century and more, speeds have increased, average times have decreased and the sport has been encompassed by a vast array of technologies, from the standardization and improvement of track surfaces, to the global commodification of the running shoe, and changing technologies for timing races. The practice of running is very much an achievement of the hybrid, trained and equipped runner.

In Chapter 4 J.D. Dewsbury discusses the ephemeral embodied mobile practices of dance, and he provides a critical intervention into recent geographical debates about non-representational theory, representation and dance. Dewsbury draws upon the writings of philosophers Alain Badiou and Gilles Deleuze to show how an interrogation of the showing of dance can enable geographers to explore ontological dimensions of moving bodies, and he does this by tracing seven elements or stages in the performance of dance. At the heart of Dewsbury's chapter, is the focus on the 'philosophical show of dance'. He argues for an analysis of the movement of dance that is beyond the issue of historical (or any other) context. In this chapter mobility (as dance) is a spectacle and an event.

In Chapter 5 Eric Laurier discusses the practices of driving and, like Dewsbury, his aim is to make a critical intervention into geographic debates about non-representational theory. Drawing upon extensive ethno-methodological research and the writings of post-cognitive psychologists, Laurier questions Nigel Thrift's distinction between 'precognition' and 'cognition', and particularly his assertion that driving is frequently practised in unconscious, automatic or non-cognitive ways. In contrast, Laurier's studies of the ordinary practices of driving and passengering accounts for the everyday actions of driving without resorting to cognitive psychology or a Cartesian split between cognition and pre-cognition.

In Chapter 6 Dydia DeLyser examines flying, focusing on the practices of pioneering American women aviators in the 1920s and 1930s. DeLyser draws on the experiences of three crusading women pilots – Amelia Earhart, Ruth Nichols and Louise Thaden. She utilizes feminist understandings of embodiment and theories of practice to show how their embodied flying practices were used intentionally to challenge dominant gendered narratives about the place of women in American society and the lived spatialities that accompanied them.

Spaces

Advocates of the 'new mobilities paradigm' have consistently noted the need to consider mobilities alongside 'moorings' (Hannam et al. 2006; Crang 2002). Aircraft need airports, cars need places to park (they spend most of their time parked) and refuel, ships need ports, and we all need moments and spaces of rest. There is a long tradition, in geography, of considering the infrastructure of transport (and other forms of mobility) in both quantitative and qualitative modes (Appleton 1962; Horvath 1974; Jackson 1997; Graham and Marvin 2001; White and Senior 1983). As well as spaces of rest, or mooring, mobilities also need spaces in which to enact mobility – roads, the air, the sea, railway lines, bridges. These spaces have their own grammar which can direct or limit mobility. They produce structural or infrastructural contexts for the practising of mobility. They are agents in the production of mobilities. But spaces are not simply contexts, they are also actively produced by the act of moving. Streets, parks or cities are created through variations of what David Seamon referred to as 'place-ballets' (Seamon 1980). Practices of mobility animate and co-produce spaces, places and landscapes. While there is a long tradition of associating practices and spaces of mobility with notions of 'abstract space', 'placelessness' and 'non-places' (Jacobs 1961; Relph 1976; Lefebvre 1991; Augé 1995), we would argue that places and landscapes are continually practised and performed through the movement and enfolding of a myriad of people and things. Rather than think of places or landscapes as settings, surfaces or contained spaces through and across which things move, it is perhaps more useful to think about the ongoing processes of 'spacing', 'placing' and 'landscaping' through which the world is shaped and formed (Merriman 2004; Merriman et al. 2008). Space, place and landscape are best approached as 'verbs' rather than as 'nouns' (cf. Mitchell 1994; Cresswell 2003; Merriman 2009).

The authors writing in this section examine five spaces which occupy very different positions within networks of mobilities: roads; bridges; airports;

immigration stations; and cities. Airports, bridges, roads, and immigration stations appear to have a fairly singular function, while cities are large-scale multi-functional agglomerations of materialities and mobilities (Amin and Thrift 2002; Latham and McCormack 2004). And yet, airports vary in size and function, ranging from rural aerodromes to international airports and military airbases; facilitating business travel, holiday flights, amateur aviation, military bombing campaigns, and extraordinary rendition. As Ulf Strohmayer shows in Chapter 8, bridges have served as spaces for living and trading, as well as strategic and symbolic crossing points. Streets and roads have served as spaces of carnival, protest, surveillance and play, as well as spaces for vehicular and animal movement (Fyfe 1998). Modern day immigration stations continue to process refugees and asylum seekers, while former immigration stations serve a very different function after their renovation as museums that celebrate and commemorate the heritage of immigrant mobilities. The chapters in this section examine spaces which are associated with different temporalities and durations of dwelling, and they trace the geographies of spaces designed for different constituent publics. Immigration stations are designed to arrest and control movement, processing people according to their past and potential future mobilities, and often confining them for lengthy periods of time. Cities are frequently associated with speed, movement, energy, and a 24/7 economy and culture, but they are also spaces of continuous dwelling and of innumerable fixities. What's more, we could just as easily have included a chapter on 'villages' or 'the countryside', which are far from being places of fixity. Streets and roads are upheld as democratic public spaces of social interaction, but anti-car protestors have argued that the increasing dominance of the motor car has resulted in the corrosive privatization of public space (Merriman 2007). Airports have been likened to large towns or small cities, which employ thousands of people as well serve their flying publics.

In Chapter 7, Peter Merriman examines the spaces of the road, focusing on the work of American landscape architect Lawrence Halprin in envisioning and scoring US streets and freeways in the 1960s. Merriman discusses Halprin's collaborations with his wife, the pioneering dancer Anna Halprin, and he examines how he drew upon ideas of movement, choreography, notation, and embodiment from dance in an attempt to understand and shape people's more-than-representational practices of inhabiting landscapes (including roads). He discusses Halprin's involvement in drafting US highway design principles and examines his scores for streets, before highlighting how Halprin's explorations of landscape architecture and dance choreography represent an attempt to engineer the affective potential of driving spaces.

In Chapter 8 Ulf Strohmayer discusses the spaces of bridges, examining how these vital structures can be positioned at the intersection of geographical writings on mobility and architecture. Drawing upon historical research on European bridges from the medieval era to the present day, he shows how bridges are not singular or homogenous structures, but they vary immensely in their design and function. Bridges are a very particular type of space, where one is between places and forced to make decisions, and by way of example Strohmayer discusses the historical geographies of the Pont Neuf and Pont-au-Change in Paris.

Another kind of space in which mobility is conditioned, and where mobility merges with architecture, is the airport. In Chapter 9, Peter Adey discusses this iconic space of modern mobility. Adey approaches the airport by way of a discussion of vectors, which he take to be 'embodied and experiential', and he traces the ways in which people, things and airports may be considered as vectors – as 'path-like'. Adey's aversion to pointillist thinking results in an approach to airports that is underpinned by a post-structuralist sensitivity to the incessant mobilities of the world.

In Chapter 10, Gareth Hoskins and Jo Maddern transport us to the spaces of two US immigration stations: Angel Island in San Francisco and Ellis Island in New York. They examine how, in the late nineteenth and early twentieth centuries, Angel Island and Ellis Island were important sites for the processing and regulation of 1 million Asian and 12 million European immigrants, respectively, where they were assessed, categorized and either admitted or excluded. Hoskins and Maddern then proceed to examine the revisioning of these immigration stations as sites of commemoration; as museums presenting stories today about the significance of mobility and immigration to US identity, history and heritage.

In Chapter 11, David Pinder discusses the space of the city – perhaps the exemplary kind of modern mobile space. His focus is on a series of conceptual architectural projects by European avant-garde architects from the 1950s and 1960s, in which cities were envisioned as moving, machinic agglomerations of floating, modular parts. Cities, here, are not just cross-cut by movements, rather they are envisioned as shifting, moving, roaming entities, and Pinder discusses at length a number of projects that were associated with the Archigram group of English architects. In addition to the creativity of these projects, Pinder also delineates an implicit politics of mobility.

Subjects

There is an important history and geography of how particular means and styles of moving have come to be associated with distinctive subject positions. The subject 'citizen', for instance, has been defined as much by the right to move as by the particular cities or nations to which he or she belongs (Cresswell 2009). Familiar figures such as the tourist and the refugee are also defined through their mobilities and are subjects of relatively recent origin (Bauman 1993). They are mobile and modern despite their different practices and experiences. These subjects are defined by representational schemes that lie beyond the scale of the individual. They have been constructed and represented in law, newspaper accounts, novels and films, but these subject positions are also inhabited, resisted and manipulated through practice. Representations of particular subject positions frequently caricature such figures, stereotyping the manner in which people of different gender, class, ethnicity, wealth, age, sexuality or nationality are expected to occupy particular mobile subject positions, and erasing the differences of those same individuals. Both habits are evident in representations of the five subject positions discussed in this section: commuter; tourist; migrant worker; vagabond; and refugee. Migrants (in the developed world) are regularly represented by tabloid media as poor, uneducated, non-white and as a drain on state services, but such discourses overlook not only the complex histories and geographies of migration, but also the diverse array of individuals who have assumed this subject position, from those internal migrants whose movements underpinned the industrial revolution in Britain in the eighteenth and early nineteenth centuries, to wealthy educated bankers moving between global financial centres, and undocumented migrants working for low wages in sweatshops that underpin the production of globalized 'Western' brands. Likewise, the qualitative experiences of commuting have rarely been analysed by academics, and yet they have been the subject of biting satires and comedies which, in Britain, have frequently focused their attention on the disillusioned middle-class male commuter – Tim Edensor discusses the 1970s sit-com *The Fall and Rise of Reginald Perrin*; to which we might add Tom Good from *The Good Life*, who gives up his job and daily commute, to pursue a self-sufficient lifestyle with his wife Barbara in their suburban home and garden. In such representations of mobile subjectivities – ranging from the intentionally witty and satirical, to the exclusionary and xenophobic – the individual experiences of those inhabiting such subject positions frequently get overlooked and erased, but it is also evident that individuals practice and inhabit these subject-positions in many different ways – challenging and reworking conventional caricatures and stereotypes ... moving differently.

In Chapter 12 Tim Edensor discusses the figure of the commuter. Drawing upon Henri Lefebvre's final book *Rhythmanalysis* (2004), Edensor traces 'the spatial and experiential dimensions of commuting rhythms', analysing extracts from the diary of a young female commuter who makes a daily commute by car in North-West England. Edensor discusses the pleasurable sensations and the predictable drags of commuting, and he illustrates his arguments by focusing on four key themes: synchronicity, consistency, disruption and sensation. These observations serve to question the familiar stereotypes of the commuter as a bored character inhabiting 'dead time'.

In Chapter 13, Mike Crang focuses on the subject position of the tourist. In a highly reflexive, confessional ethnography of his trip to the Greek island of Kefalonia, Crang reveals the complex mobilizations of the tourist destination, tourists themselves, and of his own academic mobilities as a researcher. He examines how Kefalonian tourism has become associated with the film based on Louis de Bernières' best-selling novel *Captain Corelli's Mandolin* (1994), which was set and filmed on the island. Crang explores his complex position as both an academic researcher and tourist, examining multiple representations of Kefalonia, and describing the many encounters with his tourist research participants, some of whom are tracking down the sites where filming took place.

While the tourist and the academic are relatively privileged mobile subjects, migrant workers often inhabit the opposite end of the spectrum. In Chapter 14, Elizabeth Lee and Geraldine Pratt discuss the lives of two migrant women – Mexican janitor Faviana, who lives in the USA, and Filipina care worker Liberty, who lives in Canada – revealing how they have quite different (yet in some ways similar) experiences of migration policies in two affluent North American countries. Lee and Pratt show how through the act of migration both women have been forced to work in low-skilled 'feminised' jobs, separated from their children, and caught up in an ongoing situation of vulnerability and uncertainty. This is a long way from the world of tourists in Kefalonia.

In Chapter 15, Tim Cresswell focuses on the mobile figure of the vagabond or vagrant. In an account which traces the 'curious career' of the vagabond figure from the medieval era through to the present day, Cresswell traces the role of the vagabond in law, literature, art, popular music and theory. In doing so he shows how this figure has been variously constructed as 'hero and villain', as 'threat and salvation', with his mobility being legislated against, romanticized, and upheld as a postmodern figure *par excellence*. He also alerts us to the necessity of seeing the mobilities of the past in the mobilities of the present and future.

The refugee may be the modern incarnation of the vagabond. In Chapter 16, Alison Mountz examines the mobilities of the refugee. Drawing upon Giorgio Agamben's writings on the paradoxical spaces of sovereign power, she shows how as sovereign territory has become dispersed, so the categories of refugee and migrant have become increasingly blurred, resulting in caricatures and conflations of the two subject positions by sensationalist newspapers and an increasing policing of borders and mobile subjects by state authorities.

Moving On

This book is wilfully wide-ranging in terms of the theoretical approaches taken and the types of mobility considered. Some chapters are historical, while others are contemporary. All, however, remain focused on the ways in which bodies and things move, the political, cultural and aesthetic implications and resonances of these movements, the meanings ascribed to these movements, and the embodied experiences of mobility. We have deliberately encouraged the authors to cover a diversity of scales of movement (from the steps of the walker to the inter-continental migrations of the refugee) and to address the histories as well as the geographies of mobility (from the medieval tramp and bridge, to contemporary migrant workers and museums). It is through this wilful diversity that the central theme of mobility emerges most strongly. It is also through this eclecticism that we believe it is possible to show how a 'mobility turn' in human geography is different from other kinds of mobile explorations (outlined above) that have peppered the discipline's history. This approach, we believe, allows links to be made between transport and dance, or migration theory and tourism. Occasionally the juxtapositions can be jarring and possibly even disturbing (such as that between a touristic/ethnographic exploration of Kefalonia and the migrant experiences of domestic workers in the US and Canada). But in addition to highlighting the variety of mobile experiences that characterize a world in motion (a world too often generalized and homogenized as simply 'mobile') they also point to some of the connections and logics that link the seemingly disparate worlds of the metaphorical 'tourists' and 'vagabonds' (Bauman 1998).

References

Abler, R., Adams, J. and Gould, P. (1971), *Spatial Organization: The Geographer's View of the World* (Englewood Cliffs, NJ: Prentice Hall).

Adey, P. (2009), *Mobility* (London: Routledge).

Allen, J. and Pryke, M. (1999), 'Money cultures after Georg Simmel: mobility, movement and identity', *Environment and Planning D: Society and Space* 17, 51–68.

Amin, A. and Thrift, N. (2002), *Cities* (Cambridge: Polity).

Appleton, J.H. (1962), *The Geography of Communications in Great Britain* (London: Oxford University Press).

Augé, M. (1995), *Non-Places: Introduction to an Anthropology of Supermodernity* (London: Verso).

Bauman, Z. (1993), *Postmodern Ethics* (Oxford: Blackwell).

Bauman, Z. (1998), *Globalization: the Human Consequences* (New York: Columbia University Press).

Bissell, D. (2010), 'Passenger mobilities: affective atmospheres and the sociality of public transport', *Environment and Planning D: Society and Space* 28, 270–89.

Blunt, A. (1994), *Travel, Gender and Imperialism* (London: Guilford).

Boyle, P.J. and Halfacree, K. (1998), *Migration into Rural Areas: Theories and Issues* (Chichester, UK: Wiley).

Bull, M. (2000), *Sounding out the City: Personal Stereos and the Management of Everyday Life* (Oxford: Berg).

Bunge, W.W. (1966), *Theoretical Geography* (Lund: Royal University).

de Certeau, M. (1984), *The Practice of Everyday Life* (Berkeley, CA: University of California Press).

Clifford, J. (1997), *Routes: Travel and Translation in the Later Twentieth Century* (Cambridge, MA: Harvard University Press).

Cook, I. et al. (2004), 'Follow the thing: Papaya', *Antipode* 36, 642–64.

Crang, M. (2002), 'Between Places: Producing Hubs, Flows, and Networks', *Environment and Planning A* 34, 569–74.

Cresswell, T. (2003), 'Landscape and the obliteration of practice', in K. Anderson, M. Domosh, S. Pile and N. Thrift (eds), *Handbook of Cultural Geography* (London: Sage), 269–81.

Cresswell, T. (2006), *On the Move: Mobility in the Modern Western World* (New York: Routledge).

Cresswell, T. (2009), 'The Prosthetic Citizen: New Geographies of Citizenship', *Political Power and Social Theory* 20, 259–73.

Crouch, D. (ed.) (1999), *Leisure/Tourism Geographies: Practices and Geographical Knowledge* (London: Routledge).

Crowe, P.R. (1938), 'On Progress in Geography', *Scottish Geographical Magazine* 54(1): 1–18.

Domosh, M. (1991), 'Toward a Feminist Historiography of Geography', *Transactions of the Institute of British Geographers* 16, 95–104.

Driver, F. (1999), *Geography Militant: Cultures of Exploration in the Age of Empire* (Oxford: Blackwell).

Duncan, J. and Gregory, D. (eds) (1999), *Writes of Passage: Reading Travel Writing* (London: Routledge).

Freeman, M. (1999), *Railways and the Victorian Imagination* (London: Yale University Press).

Freeman, M. and Aldcroft, D. (1985), *The Atlas of British Railway History* (London: Croom Helm).

Fyfe, N. (ed.) (1998), *Images of the Street* (London: Routledge).

Graham, S. and Marvin, S. (2001), *Splintering Urbanism: Networked Infrastructures, Technological Mobilities and the Urban Condition* (London: Routledge).

Hägerstrand, T. and Pred, A.R. (1981), *Space and Time in Geography: Essays Dedicated to Torsten Hägerstrand* (Lund: CWK Gleerup).

Hannam, K., Sheller, M. and Urry, J. (2006), 'Mobilities, Immobilities and Moorings', *Mobilities* 1, 1–22.

Hanson, S. and Pratt, G.J. (1995), *Gender, Work, and Space* (London: Routledge).

Harvey, D. (1969), *Explanation in Geography* (London: Edward Arnold).

Horvath, R.J. (1974), 'Machine space', *The Geographical Review* 64, 167–88.

Jackson, J.B. (1997), *Landscape in Sight: Looking at America* (New Haven, CT: Yale University Press).

Jacobs, J. (1961 [1965]), *The Death and Life of Great American Cities* (Harmondsworth: Pelican).

Kaufmann, V. (2002), *Re-thinking Mobility: Contemporary Sociology* (Aldershot: Ashgate).

Latham, A. and McCormack, D.P. (2004), 'Moving cities: rethinking the materialities of urban geographies', *Progress in Human Geography* 28, 701–24.

Laurier, E. and Brown, B. (2008), 'Rotating maps and readers: praxiological aspects of alignment and orientation', *Transactions of the Institute of British Geographers* 33, 201–21.

Laurier, E., Lorimer, H., Brown, B., Jones, O., Juhlin, O., Noble, A., Perry, M., Pica, D., Sormani, P., Strebel, I., Swan, L., Taylor, A.S., Watts, L. and Weilenmann,

A. (2008), 'Driving and passengering: notes on the ordinary organisation of car travel', *Mobilities* 3, 1–23.

Law, R. (1999), 'Beyond "Women and Transport": Towards New Geographies of Gender and Daily Mobility', *Progress in Human Geography* 23, 567–88.

Lefebvre, H. (1991), *The Production of Space* (Oxford: Blackwell).

Lefebvre, H. (2004), *Rhythmanalysis: Space, Time and Everyday Life* (London: Continuum).

Matless, D. (1998), *Landscape and Englishness* (London: Reaktion).

Merriman, P. (2004), 'Driving places: Marc Augé, non-places and the geographies of England's M1 motorway', *Theory, Culture, and Society* 21(4–5): 145–67.

Merriman, P. (2007), *Driving Spaces: a Cultural-Historical Geography of England's M1 Motorway* (Oxford: Wiley-Blackwell).

Merriman, P. (2009), 'Mobility', in R. Kitchin and N. Thrift (eds), *International Encyclopedia of Human Geography (Volume 7)* (London: Elsevier), 134–43.

Merriman, P., Revill, G., Cresswell, T., Lorimer, H., Matless, D., Rose, G. and Wylie, J. (2008), 'Landscape, mobility and practice', *Social and Cultural Geography* 9, 191–212.

Michael, M. (2000), *Reconnecting Culture, Technology and Nature: from Society to Heterogeneity* (London: Routledge).

Mitchell, W.J.T. (1994), 'Introduction', in W.J.T. Mitchell (ed.), *Landscape and Power* (London: University of Chicago Press), 1–4.

Nash, C. (2000), 'Performativity in practice', *Progress in Human Geography* 24, 653–64.

Pirie, G. (2009), *Air Empire: British Imperial Civil Aviation, 1919–39* (Manchester: Manchester University Press).

Pred, A. R. (1977), 'The choreography of existence: comments on Hägerstrand's time-geography and its usefulness', *Economic Geography* 53, 207–21.

Relph, E. (1976), *Place and Placelessness* (London: Pion).

Seamon, D. (1979), *A Geography of the Lifeworld: Movement, Rest, and Encounter* (New York: St Martin's Press).

Seamon, D. (1980), 'Body-Subject, Time-Space Routines, and Place-Ballets', in A. Buttimer and D. Seamon (eds), *The Human Experience of Space and Place* (London: Croom Helm), 148–65.

Sheller, M. and Urry, J. (2006), 'The new mobilities paradigm', *Environment and Planning A* 38, 207–26.

Stoddart, D. (1986), *On Geography and its History* (Oxford: Blackwell).

Thrift, N. (1990), Transport and communication 1730–1914, in R.A. Dodgshon and R.A. Butlin (eds) *An Historical Geography of England and Wales*, 2nd Edition (London: Academic Press), 453–86.

Thrift, N. (1996), *Spatial Formations* (London: Sage).

Thrift, N. (2000), 'Non-representational theory', in R.J. Johnston, D. Gregory, G. Pratt and M. Watts (eds), *The Dictionary of Human Geography*, 4th Edition (Oxford: Blackwell), 556.

Urry, J. (2000), *Sociology Beyond Societies: Mobilities for the Twenty-First Century* (London: Routledge).

Urry, J. (2007), *Mobilities* (Cambridge: Polity).

White, H.P. and Senior, M.L. (1983), *Transport Geography* (London: Longman).

PART I
Practices

Chapter 2
Walking: New Forms and Spaces for Studies of Pedestrianism

Hayden Lorimer

Introduction

Walking is a social practice that has been subject to increased academic scrutiny during the past decade. Amongst a variety of researchers – most notably cultural geographers and social anthropologists – new 'walking studies' have focused attention on what might be considered the founding, or constituent, elements of this most basic of human activities, namely: the *walk*, as an event; the *walker*, as a human subject; and, *walking*, as an embodied act. Whether treating the *walk-event*, the *walker-person* or the *walking-act* as the starting point of analysis, these studies commonly figure pedestrianism as a *practice* (Lee and Ingold 2006). More than this, the lived, or practised, realities of walking are understood variously, as being: reflective of changing social forms and norms (Edensor 2000); expressive of diverse cultural meanings (Lorimer and Lund 2008); and, leaving distinct impressions, both corporeal and materially substantive (Michael 2000). This current preference for a cultural interpretive frame in walking studies (Olwig 2008) contrasts with the earlier, fairly slim, treatment of the subject in social science research where the primary significance afforded walking was as the locomotive means to very particular ends. Walking was destination-oriented, generally regarded a functional mode of transport, and shaped by economic choices and constraints. Its study was therefore one of time-in-motion-across-space, and determined a means to enable accurate predictive modelling and mapping of human behaviour and associated environmental preferences. The current preference for a cultural-interpretive mode of description and analysis is different once again from histories of past pedestrian cultures originating in the humanities generally taking a lead from the literary, poetic or artistic *representation* of lives spent travelling or journeying on foot (Bate 1991, 2000; Landry 2001; Wallace 1993; Taplin 1984; Solnit 2000). Here, contrasting versions of the peripatetic lifestyles of travellers and their shaping of landscapes are channelled through a history of greater aesthetic movements abroad in society. Thus, we find ways of walking explained according to the expressive work of Romantic poets, the picturesque tradition in landscape painting and experimental writers chronicling their experiences of the metropolis at the *fin-de-siecle*. Working in this tradition, the walker-writer Robert Macfarlane has attempted to classify those characters in whom he finds peripatetic inspiration:

> The category "walker" has many subdivisions. There are the marathon men: the long-distance land artist Richard Long for instance, or Thomas Coryat, who in 1612 marched from London to Agra, and who liked to refer to himself as propatetique (that is, "a walker forwarde on fette") rather than a peripatetique (that is, a person "who meerely walks arounde"). There are the flaneurs: De Quincey, Defoe, Sean Borodale – anthropologists of the street, botanists of the asphalt, prying their way round cities and towns. There are the psycho-geographers, the downriver dowsers – Iain Sinclair, Chris Petit, Will Self. There are the adventurers – Robert Louis Stevenson, Stephen Graham. And there are the wanderer-wonderers – Samuel Taylor Coleridge, Virginia Woolf. (Macfarlane 2007, 79)

Macfarlane's taxonomy is not exhaustive. Such has been the level of interest in exploring and explaining different cultures of walking that the contemporary body of walking-work demands a more detailed, and suggestively systematic, typology. This definitional task, contributing to a greater configuration of mobile geographies, would need to explore the different sorts of cultural resonance that walking can have, *and*, consider the approaches taken in research making this social practice an identifiable subject of study. What I propose is doubtless similarly incomplete, and in places impressionistic, since cultural interest and research activity has clustered around particular fields of concern. If what it amounts to is a miscellany of walking studies – past, present and potential – then crucially it is one that seeks to understand the geographical dimensions, and inflections, of respective inquiries. In what follows, I have chosen to group interests under four thematic headings: first, walks as the product of places; second, walks as an ordinary feature of everyday life; third, the reflections of the self-centred walker, and fourth, walkers who are wilful and artful. A critical, synthetic consideration of these four thematic interests comprises the first part of this chapter. The chapter's second part seeks to mobilise some key contentions from such 'new walking studies' by considering the part they play in the situated and specific context of walkers' experiences of passage in hills and mountains.

A Miscellany of Walking Studies

Walks as the Product of Places

According to some, a walk can be understood as a cultural activity that is made distinctive and meaningful by the physical features and material textures of place. Some walks take shape through the collective observance of a regional tradition or custom, coupled closely with the local lie of the land. In this sense, traditional or commemorative walks are oriented according to prominent landscape features, and arranged by particular kinds of topographical association. Specific routes – taking in key landmarks or offering long prospects – are followed on special days of the calendar. Hilltop constructions (cairns, beacons, crosses or monuments) and their elevated viewpoints are often the defining features of such walks. Or, we might usefully think of local festivals of longstanding where the 'beating of bounds' (or the determination of rights of way) demands that, in their movements, a collective of walkers circumnavigate the historical limits of a settlement or commons, and symbolically stake out both its, and their, territorial claims. The current vogue for establishing new walking festivals – part of the place promotion strategies adopted in numerous settlements across rural Britain – recreates such civic traditions anew. Here, recognisable routes are followed by walking groups, normally on an annual basis, as a means to renew bonds with a place, and, with a community of fellow walkers. Such site-specific geographies of walking can be linear and progressive, as well as nodal or circulatory. On the long-distance paths along which walkers travel for several days, continuously moving 'from A towards B', routes are suggested by the long-term processes of evolution that shape coastal margins and cleave open mountainous landscapes. By its natural inclinations, topography offers the path most obvious, or most comforting, for safe *and* scenic passage. Such routes are attractive for solo walking, or in the company of companions.

In acts of pilgrimage, the point-to-point walk takes sacred expression, existing as a line in the landscape shaped by common observance of a religion. Pedestrianism born along by a shared faith affords a sacred quality to specific routes and terrains and, as Candy argues, it has been for centuries to 'encounter places that inspire sensations, fears, associations, memories or fervour' (2004, 16). The ritualised (and sometimes repeat) passage of Christian pilgrims represents a statement of self-sacrifice and, when sequenced according to the stations of the cross, links together a series of points and places (Lund 2008). Such devotional walking acts are regarded as an intensifier of life as on-going process, and are not celebrated so narrowly as to only acknowledge the accomplishment of distance covered, but of crossing points, transitions and thresholds en route where practitioners can claim that they got to know their self and their spirituality better (Coleman and Eade 2004).

Walks as an Ordinary Feature of Everyday Life

While certain walks can be followed as a customary, religious or festive practice (and are thereby afforded greater significance by the relative infrequency of their observance) others are undertaken habitually, and in some cases, daily. Often these types of walk cover short-distances and are task-centred or goal-oriented; though they are no less social for their repetitive nature or functional purpose. Often these walks are made alone, though sometimes in the company of companions or dependents. Consider the dog-walkers who visit the local park twice daily to ensure that they themselves *as well as* the family pet are kept well exercised; or, the early walk made to the neighbourhood convenience store to buy a morning paper; or, the walk to school, shared by parents (or carers) and a gaggle of children; and finally, the commute-on-foot to a place of work, and later home again, following a regular route, perhaps part-way with a colleague, undertaken at the same times each day. That each is recognisably routine in quality goes some way toward explaining why such walks are favoured as the setting for observational experiments in rhythm-analysis (Mels 2004). Such analyses configure the massing of people and things as a mobile assemblage, constituted of trans-personal flows of movement, social relations and embodied associations (Anderson 2004; Lee Vergunst 2008). Closely related, but different in founding rationale, ethno-methodologists have considered the mechanics of pedestrians negotiating two-way passage along a crowded pavement as formative of the ordinary social realm (Wolff 1973).

Walking *to* or *from* work is of course an entirely different sort of undertaking than walking *as* work. Changing labour practices – and a reliance on the personal computer as the primary interface for *doing* work in post-productive, professionalised, service-sector economies – mean that occupational sedentarism is increasingly the norm among today's workforces. Here, it is worth momentarily reflecting on personal experience. This chapter is a piece of writing on the subject of walking which was composed in a static, sedentary position. Those sectors of employment where walking remains central to work seem increasingly the stuff of an older society, now passing away, predominantly agrarian in organisation, and outdoors in its operations. Think here of the shepherd, the ploughman, the postman, the paperboy, even the salesman peddling wares door-to-door. Once, itinerant or migrant farm workers were dependent on their feet to get them between seasonal jobs. However, a world of foot-work should not be too quickly dismissed. It does remain a feature of many working lives, and part of tending for others' food, safety, pets or data. For the staff-nurse, the security guard and the household cleaner, mobile geographies of walking are an essential means of 'getting about' so that the prescribed tasks of the job can be done. Still greater care is necessary in mapping the ordinary sociality of walking. Although working lives might seem at first wholly desk-based, when examined empirically and at the micro-scale, walking is the commonest of relational and embodied working practices. To walk indoors – we should not forget – is to join up the most necessary bodily functions with mundane employment tasks and social interactions. And of course, on deeper reflection, many of the ideas advanced and observations made in this chapter took shape as I travelled about on foot; between my desk and the Departmental Office, and the nearby catering facility to buy a cup of coffee.

The populations of advanced economies notwithstanding, there remain among indigenous peoples worldwide a great diversity of subsistence communities for whom walking is a way of living, rather than a means to *make* a living. The rupture caused by the industrial revolution – 'when walking ceased to be part of the continuum of experience and instead became something consciously chosen.' (Solnit 2000, 265) – was not experienced universally or evenly. For indigenous groups the walking act is central to existence, and a continuous means to learn and to teach place-knowledge: thus, living and walking exist along the same continuum (Ingold 2004, 2007; Legat 2008; Tuck-Po 2008).

The Self-centred Walker

For some practitioners and researchers, walking is a compelling means to find a better sort of fit between self and world. Walking offers an embodied space where searching questions are considered, and sometimes answered. Thus, for the walker in philosophical mood, queries such as 'where do I stand?' 'how have I come this far?', and 'have I gone far enough yet?' work on a spiritual plane, rather than a practical-referential one. Answers take ontological shape and meditative form. To consider matters of being, and becoming, whilst on the move is to fold walking into the quest for greater harmony, and to meet philosophical needs that are both deeper and wider set. Indeed, it is sometimes rhythms and feelings understood to be produced *in* the act of walking that lead to feelings of contentedness or wholeness. This is a continuation of a 'long-walking' pedestrian tradition. All sorts of virtue are commonly said to accrue from walking, happiness being the most fundamental, but evenness of temperament and emotional balance rank highly too. By an earlier, thin-lipped and ruddy-cheeked sort of credo, the pastime of walking offered to its practitioners a moral compass by which habits of right living were forged. Cultural outcrops from a British history of outdoor pursuit show the popular personification of self-improvement and self-discipline in the spry, shirt-sleeved figure of the hiker (Matless 1998).

Walking as an activity upon which founding aspects of the self can be centred finds contemporary expression in ideas of release, renewal and replenishment. Such affective-emotional geographies are now an emerging topic of inquiry in walking studies. In the midst of walking it is possible to think about being-in-the-world, and find grounds for a freer experiential exploration of what it is *to be* or *become* (Wylie 2002, 2005, 2007). Here cultural analysis turns inwards – seeking to access walkers' intimacies of encounter – rather than focusing on the outer (symbolic or socialised) meaning of walking acts. This mode of research demands a critical sort of self-regard as research practice, and a willingness to accept subjectivity formed from auto-ethnographic encounters with affective landscapes. The approach taken is at once a very personal and self-conscious form of accounting, and simultaneously, a chronicling of the pure perception of worldly phenomena. Passages narrated in the first person singular about the experience of walking open up spaces for enquiries into the limits of the visual, the physical and the representational. Wylie has termed this a 'post-phenomenological' approach where self and landscape are always emergent, constantly shifting through repertoires of the unbidden, of affective and kinaesthetic contact, and then dissipating just as easily. 'Self-identity' does not do proper justice to what is encountered by the reflexive researcher-subject. The discovery of kinds of sensational attunement – emanating from the body, responding to circulations of forces and phenomena – demands more than an axis drawn between soles and soul. Becoming happens, on foot, in a greater and more diffuse field of about-ness than the individual can ever encompass. Hereabouts, the walker is both visceral presence and will-o'-the-wisp. Meanwhile the topographical forms of the places traversed do cohere as coastline, tor or hill-slope, Wylie's encounters are occasioned through forms, depths and fields that constitute a sensorium of experience. Acuity and awareness surface in moments, or longer spells. Receptiveness is explained in terms of joy, or pain, or fatigue, or love (Wylie 2005).

This mode of walking as self-searching can also take expression according to a therapeutic vocabulary, where well-being is a threshold to be crossed, an equilibrium reached, or a state found, spiralling out of matters of mind, body and spirit. Walking has come to be figured an effective means of self-help, a journey to improved mental health, and an acceptable way to get to know ourselves better (Conradson 2005). Mobility is what produces stability, or, a greater acceptance of internal instability. Of course, the spacing of subjectivity is not always so well sealed. Studies which place the self – or others' selves – centre-stage must also be attentive to sensory and appreciative relationships with the environment. Among such communities of appreciation, to walk is to feel oneself engaged in a sustaining conversation with landscapes and to seek guidance from the natural world. Hereabouts, currents of spiritualism, animism and nature worship spring forth, braiding and blending together differently. Figured so, walking can become a defining expression of a wider search for alternative ways of living; 'the good life', placed in opposition to contemporary cultures of consumption. In much the same vein, walking can become a physical and therapeutic form of weekend escapism, amidst the more conventional styling of life, and its material trappings. A walk becomes a short-lived but highly-prized reward for the longer grind and routine of work. Walking is understood to produce intense feelings of liberation or refreshment, a stronger connection with what is elemental, and a slowing in the otherwise hectic pace of life (Lorimer and Lund 2008). This re-charging can be a necessary reminder that one's body still pulses during effort, and is capable of demanding physical exertion. The lingering afterglow, and embodied ache, felt on having come home (and 'put your feet up') is one of the motivations for having gone out for a walk in the first place. The capacities and limits of physical experience that give texture to any walk can also be a means to consider aspects of human biology and social difference (Macpherson 2008). Walking does not come naturally to all, and is not embodied in exactly the same way. Walking with a disability, such as visual impairment, creates differing conditions of experience, and produces different kinds of competency, and abilities to visualise surroundings.

Walkers who are Wilful and Artful

With some justification walking can be treasured and held dear as one of the most humble of human acts. Alternatively, each footfall can be figured as an action of potential, awaiting the expression of artistic invention or political intent. When executed as a mass performance of resistance, or as a creatively orchestrated kind of individualism, walkers render a basic developmental competency as a politically wilful or deeply artful act. Thus what can be read at one level as biomechanical and functional by another becomes a complex symbolic statement.

Many contemporary walking studies personify the walker in one of two ways: either as a politically-savvy activist, or, as an artistically-inclined activist. For each 'type' of walker – or conceivably a fusion of the two – particular notice has been taken of the diverse ways that walkers actively intervene in the social life of the city or the countryside (Pinder 2005; Philips 2005). Among walker-activists – and those cultural analysts numbered among them – the focus falls on strategies for collective or community-based actions; in some cases links are made to pedestrian actions in the past, and to earlier generations of inhabitants, whose pedestrianism continues to set a telling example in the present. In such instances, the practice of walking is mobilised as a form of popular protest where the placement of feet is done most determinedly, as an act intended to have powerful symbolic effect. When galvanised, people take a walk in defense of different causes: the authentic character (or scenic appeal) of places (such as parks and gardens); the ancient rights of way enabling access on foot through cherished landscapes; and, the legitimate use of public spaces for mass gatherings and protests. Of course, historically the mass trespass and the protest march have been effective techniques for the embodied and mobile expression of claims to place and of popular opposition to being impeded or inhibited. Grassroots, campaigning-on-foot often happens as a result of local contestations. The spur to join a walk can centre on perceived threats to (feelings of) common ownership. The situated circumstances are various, though resource prospecting, the privatisation of property, encroaching commercial development and landscape disfigurement figure consistently. For the Palestinian lawyer Raja Shehadeh (2007), the continuance of a lifelong habit of hill walking around Ramallah is something done not so much in defense of a territory, since the physical changes wrought by Israeli re-settlement are irreversible, and rather is undertaken in pursuit of a vanished landscape and the few remaining relics of a cherished way of life. To follow known paths – and to write of the experience of walking them – is to give expression to his people's feelings of loss and anger, and their stoical and realistic outlook. Shehadeh's insistence on visiting old haunts by walking is a political statement, and his notes on these outings are the medium that makes an art of politics.

As an informal social movement, radical pedestrianism need not only be reactive or resistant. Site-specific community projects (for example, the extension or decoration of path networks, and the creation of sculpture gardens, community gardens and woodland walkways) are celebrations of place-attachment and place-memory designed specifically to be enjoyed on foot (Morris and Cant 2006; Butler 2006). In a British context, the flourishing of these kinds of local cultural initiative has enabled social researchers to undertake spatially-sensitive analyses of the processes and practices by which the identities of places and inhabitants are co-constituted (Mackenzie 2006). In the same context of land-based activism, voluntary organisations such as Common Ground (in England), Duchtas (in Scotland) and the Ramblers Association (UK), are notable for having cultivated a consensual kind of pedestrian politics, while simultaneously indulging a potentially insular strain of localism.

The choreography and the performance of walking, and a diaristic chronicling of passage through an external environment, has also become a highly popular form of critical arts practice (Gooding 2004; Heddon 2008). That *the walk* has gained such widespread recognition as art medium, and art form, is attributable jointly to the 'footworks' of Richard Long and Hamish Fulton. Long first gained renown through the repeated act of walking in place to produce pieces of land-art, a mark-making process that he documented in photographs; captured most famously in a line trodden into a meadow of summer grasses. Here, a respectful alteration of the landscape is what remains from the presence of the walker-artist (Long 2007). Fulton is recognised for work that repeatedly has taken physical form as a long-distance walk – perhaps given its impressive reach his oeuvre is better described as landscape-art – and is characterised by a lightness of presence. The original walk demands no audience, is undertaken alone, designed to leave no lasting trace of passage, and other than having been completed may well end with no discernible 'product'. The fullness of Fulton's experience is later stripped-down to a graphic or typographical summation for exhibition and interpretation. The inscription of passage exists, immaterially in place, and linguistically, in a resultant practical art form (Fulton 2000a, 2000b). The accumulation of experience is framed and arranged from surviving words: 'Walking and camping for seven days/downstream returning upstream over sections of badlands/along the north banks of the Red Deer River/tent doors facing the sunrise each day/a full moon on the seventh night/Alberta Canada July 1999'. Fulton is a purist, and minimalist, articulating both humility before nature, and an exacting high moral order through his art practice. He claims to be only a servant of the walk, and not the later artwork for which he prefers to remain unanswerable. In so doing he re-defines the role of journeyman walker.

The influence of Long and Fulton can be found in so much contemporary performance art that takes the walk as its defining feature, and in so much land-art that invites the visitor to become a walker so as to better appreciate form, shape and find meaning. The walking tradition in contemporary art has been adapted to suit more inclusive forms of arts practice where different communities of practitioners convene through shared interest and collaborative endeavour. If once the land artist was a lonely and isolated individual, now, having greatly increased in number, they are commissioned by communities to give expression to shared feelings of belonging. The homely tradition of the walking-artist has been taken to a multitude of urban and rural settings and has been subject to waves of academic critique and interpretation (Stacey 2000). Indeed, when the activist walker and the artistic walker fuse to create artful kinds of social intervention, this holds greatest appeal for academic researchers. However, in much social science commentary there remains a tendency to treat walking as an undifferentiated act, which can then be read as expressive of different social identities. It is the work of performance-walkers that provides necessary reminders that there are many different styles and techniques of putting one foot in front of the other, each one charging movement with a very different voltage. Walking can be embodied in a variety of styles, or as demonstrative of personal traits: to drove with animals is not the same as to yomp with a battalion; to sleepwalk, stumble or tip-toe is to alter at a fundamental level the nature of physical movement (Whitehead 2006). Anyone unfortunate enough to have been 'frogmarched' will know this to be a markedly different walking experience than it is to saunter awhile.

As this four-part survey of recent walking studies serves to illustrate, depending on the approach taken, pedestrian cultures can be variously arranged, and different questions posed of any such formulation. And yet, before any kind of walk or walker, seems to rest a greater question, concerned with motive and value: 'What does it mean to walk with, or without, settled purpose?' The chapter's next section aims to show how finding a preferred configuration for this founding question of pedestrianism will have a telling impact on the answers found in new walking studies.

Why? Here? There? Where?

> I travel not to go anywhere, but to go. I travel for travel's sake. The great affair is to move. (Robert Louis Stevenson 1879, 46)

The compulsion to walk is all too easily confirmed as one of life's eternal mysteries if social enquiries are figured by the inquisitive researcher along the lines of '*why* do you walk?', or, alternatively, '*what* is it that you seek by walking?' Of course, the same question, with a stress placed on the disclosure of *reason* or *drive*, can be turned back on the researcher and is likely to have similar conversational effect; that of shutting things down. It is akin to asking 'Why do birds sing?'; such questions are many-times-posed, yet remain deeply unanswerable. Having a fascination for walking can therefore result in favouring the quest for underlying meaning, and thus, a temptation to avoid the sheer physicality, the actual undertaking, of the action of 'getting along'. Walking, we now understand, need not always, or only, derive function or purpose from movement or motive leading towards destination, say *between* two significant points, or, the goal of returning to the same place. Walking can be considered a mode of being that wills itself, creating purposes of its own, which emerge (and just as easily pass away) during the experience of travelling on foot. The merits of a walk can be short-lived *and* slow-burning, incidental *and* associational. Better then to ask '*what is* a walk?' Or, the same question might helpfully be refigured once again, to: '*how* do you walk?' To alter its formulation so, is to open up action and experience, becoming sensitive to the transience *and* the durability of experience. It is to make greater sense of the spaciousness of walking, allowing room for forms of meaning that are lived, and mobile, *and* to remain sensitive to the particulars of geography that a walker might still demand.

For the 'what' and 'how' questions to be posed, and remain answerable, ways of walking ought to be treated as a whole thing. In truth, the act of walking does not always disaggregate very easily though artists, photographers and performers, we should note, have sought to differently notate (and photograph) the fundamental biomechanics necessary for human locomotion (Solnit 2000, 2003). Rather than being stripped back, calcified or freeze-framed, as an abstraction or visualisation, it is preferable to regard walking as process; a continuous flow of well-orchestrated bodily movements creating the apparently simple effect of forward movement. Studying the particular ways in which walking happens amidst this flow, and between participants and environments, can combine carefully observed social interactions and individual descriptions of practice.

Bodies Walking: Abreast, In Front, Behind

Such observation – most effectively undertaken through participatory walking – enables inquiries into how community, companionship and care can form through types of passage and embodied sociality, across varied terrain and in different sorts of setting. Whether it happens in isolation or in company, in the shopping mall or on the mountainside, the practice of walking should not be considered as formless, care-less or thoughtless, but rather as a practical accomplishment formed through a flow of activity. In what follows, a series of brief embodied instances are presented for illustration and consideration.[1]

First, let us briefly consider the solo hill-walker steadily ascending, for whom choice of route seems a solitary undertaking. Yet, the footfalls of previous walkers will have already created a 'desire line' through the landscape: the force and weight of bodies, and traction of boots, wearing away ground vegetation, even creating areas of soil erosion and slippage. The desire line will have been dictated by the lay of the land; previous footprints sometimes marking the route that presents least resistance to weary legs, sometimes showing up the one offering the most direct route uphill. Visible to the eye, the desire line is an easy lure for all those who follow and are content to defer decisions on route-selection. Safety exists in numbers; even the company of those who have long since departed the scene. 'Stick to the path!' is a parental injunction, and 'Don't Stray!' a warning we learn in childhood. Footpath planners and repairers know how hard it is to dissuade adult walkers from the habit of using the well-beaten path, and to take an alternative route; for this is to resist the re-assuring lure of visible evidence. The path is a powerful trafficking device, creating a corridor of popular experience, by which the great majority of walkers come to know a mountain, or upland area, and its aspects.

Next, consider the social exchanges of a small walking party, travelling two-by-two or three abreast; the classically companionable formation of pedestrianism. For

---

1   These observations and reflections draw, in greatest part, on mobile-participatory research undertaken with walkers whose weekend hobby is to walk routes up, around, through and across hills and mountains in Scotland, and on the words of those same walkers (ESRC Ref: R000223603).

them, conversation and embodied gestures happen side-by-side rather than face-to-face. When the going underfoot is good enough the mobile-social arrangement bears useful comparison with the way that conversation happens between driver and passenger travelling by car (Laurier and Lorimer, *forthcoming*). For the walkers, mobility is similarly consequential rather than accidental. For the most part eyes are cast downward and forwards, scanning the immediate terrain to be traversed, in anticipation of sure-footed passage. And simultaneously, repeatedly, if only momentarily, faces turn inwards allowing brief moments of eye contact, or to observe facial expressions or physical gestures. But, for the greatest part, talk happens outwards, to the world. The walk shapes the rhythm of talk, and the talk shapes the rhythm of the walk. The management of speech and comportment of different bodies are thus conjoined. Social and spatial formation can also be consciously re-arranged: '…you can find three at the front going yack, yack, yack, so you drop back wanting a little silence and peace and quiet. And you go back twenty yards, you are in your own little space and the others will leave you to it'. Or physical conditions and topography enforce a change.

With a narrowing in path width, the shapely comfort of the walkers' natural arch is vitiated, and breaks. Now, where one walker leads the other will, almost certainly, follow. Conversation is produced, and embodied, differently. Comments, looks and responses must then be thrown over shoulders, and picked up by watchful eyes and attentive ears: 'it is a lot easier to walk side-by-side which makes conversation happen than if you are following a track where you shouting at the back of somebody's head'. Similarly, consider the manner in which a larger group of walkers will fall seamlessly into single file whilst ascending or descending a stretch of steeper ground. The lead walker picks a path. Those behind acceptingly take the same steps, choosing to use the exact same footholds. There is a comfort and personal security that comes in the feeling of being led, and in following. Typically, conversation will peter out on the steepest inclines and tricky stretches, beginning again only once the gradient or conditions underfoot are more forgiving. The walking party that forms a stretched out or snaking line traversing the landscape is an instantly recognisable high-level presence in (British) mountain scenery. The linear quality of the walk and of the walkers' own formation is rhythmic; encouraging participants to keep plodding onwards.

Third, consider the dynamics of a larger walking party. Short stops and starts can be taken as a collective decision, perhaps prompted by someone declaring their need for a rest, or refreshment. Or, a pause in progress can be created through the identification of a landscape prospect that demands a halt, and a corresponding desire to 'take in the view'. The satisfaction to be gained from taking in a view seems to require the stability of two feet firmly planted; possibly borne of a photographic kind of conditioning. Hilltops and mountain summits are commonly favoured and have their own particular customs and etiquette associated with visuality (Lorimer and Lund 2008). In spite of walking being one of the slowest forms of locomotion, enabling the close observation of details, sight and movement are not always commensurate or compatible. A mobile view is different – more wobbly, unset, imbalanced – than is the prospect created by stillness, and it is not accorded quite the same status. 'I can't quite take it in.', says the walker who feels compelled to press on.

Minor processes and preferences of passage such as these, punctuate and pattern the longer flow of the thing we recognise as 'a walk'. Walkers' actions can be subject to this closeness of attention, placing priority on the directly personal, unpredictable, improvised, conversational nature of much walking. But it is worth noting how – in certain settings (mountains) or under particularly testing weather conditions – walking is also disciplined, methodical, responsible, pre-planned, orderly and linear. The systematic qualities of walking are various. It is telling perhaps that qualities of improvisation are often associated with acts of urban exploration. By way of contrast, for the walker in the mountains there may be no choice of route but the one taken. Changing course may not be an option. The line taken can be *the* only line available. Routes along narrow mountain ridges, or picking a scramble towards the craggiest summits, are worn in because no (sensible) alternative exists.

Conclusion

New walking studies seem well set to continue – even to prosper – as a shared, inter-disciplinary field of concern, uniting social and geographical research and critical arts practice. This much accepted, the learning walker might usefully recognise their pedestrian practice as an on-going exercise in the tempering of expectations. Neither a walker's actions, nor answers, will explain everything about them; about whether they walk to find meaning in life, or if their life-world best fits the motif of the long walk, or morning stroll. The recent push towards a grounded consideration of walking as social practice in diverse material contexts has produced insights that speak of a physical basis in geographies of embodied presence. Future studies of walking could be as much about atmospherics as they are a world of substance: ranging from experience of place-making amidst washes of weather and elemental force fields, to swings of mood or memories that happen off-stage and away from the action. In sum, the cultural, social and political resonances that walking has been shown to have are those encompassed by the moving human body, and they must now exceed its physical form.

Acknowledgments

My thanks to Katrin Lund and Tim Ingold who collaborated on participatory research undertaken with individual hill-walkers and hill-walking clubs in Scotland; reported here, and drawn from ESRC Research Award (Ref: R000223603).

References

Anderson, J. (2004), 'Talking whilst walking: a geographical archaeology of Knowledge', *Area* 36(3): 254–61.

Bate, J. (1991), *Romantic Ecology: Wordsworth and the Environmental Tradition* (London: Routledge).

Bate, J. (2000), *The Song of the Earth* (London: Picador).

Butler, T. (2006), 'A walk of art: the potential of the sound walk as practice in cultural geography' *Social and Cultural Geography* 7(6): 889–908.

Candy, J. (2004), 'Landscape and perception: the medieval pilgrimage to Santiago de Compostela from an archaeological perspective' *e-Sharp: Journeys of Discovery* Issue 4 (University of Glasgow).

Casey, E. (1987), *Remembering – a Phenomenological Study* (Bloomington: Indiana University Press).

Coleman, S. and Eade, J. (eds) (2004), *Reframing Pilgrimage: Cultures in Motion* (London: Routledge).

Conradson, D. (2005), 'Landscape, care and the relational self: therapeutic encounters in southern England', *Health and Place* 11(4): 337–48.

Edensor, T. (2000), 'Walking in the British Countryside: reflexivity, embodied practices and ways to escape' *Body and Society* 6(3–4): 81–106.

Fulton, H. (2000a), *Magpie: Two River Walks* (Lethbridge: Southern Alberta Art Gallery).

Fulton, H. (2000b), *Wild Life: a Walk in the Cairngorms* (Edinburgh: Pocketbooks).

Gooding, M. (2004), *Artists, Land, Nature* (London: Abrams).

Heddon, D. (2008), *Autobiography and Performance* (London: Palgrave).

Ingold, T. (2004), 'Culture on the ground: the world perceived through the feet', *Journal of Material Culture* 9(3): 315–40.

Ingold, T. (2007), *Lines: a Brief History* (London: Routledge).

Ingold, T. and Lee Vergunst, J. (eds) (2008), *Ways of Walking: Ethnography and Practice on Foot* (Aldershot: Ashgate).

Landry, D. (2001), *The Invention of the Countryside: Hunting, Walking and Ecology in English Literature, 1671–1831* (New York: Palgrave).

Laurier, E. and Lorimer, H. (forthcoming), 'Other ways: landscapes of commuting' under review with *Landscape Research*.

Lee, J. and Ingold, T. (2006), 'Fieldwork on foot: perceiving, routing and socializing' in Coleman, S. and Collins, P. (eds), *Locating the Field: Space, Place and Context in Anthropology* (Oxford: Berg).

Lee Vergunst, J. (2008), 'Taking a trip and taking care in everyday life', in Ingold, T. and Lee Vergunst, J. (eds), *Ways of Walking: Ethnography and Practice on Foot* (Aldershot: Ashgate), 105–22.

Legat, A. (2008), 'Walking stories: leaving footprints', in Ingold, T. and Lee Vergunst, J. (eds), *Ways of Walking: Ethnography and Practice on Foot* (Aldershot: Ashgate), 35–50.

Long, R. (2007), *Richard Long: Walking and Marking* (Edinburgh: National Galleries of Scotland).

Lorimer, H. and Lund, K. (2003), 'Performing facts: finding a way over Scotland's mountains', in Szerszynski, B., Heim, W. and Waterton, C. (eds) (2003), *Nature Performed: Environment, Culture and Performance* (Oxford: Blackwell), 130–44.

Lorimer, H. and Lund, K. (2008), 'A collectable topography: walking, remembering and recording mountains', in Ingold, T. and Lee Vergunst, J. (eds), *Ways of Walking: Ethnography and Practice on Foot* (Aldershot: Ashgate), 185–200.

Lund, K. (2008), 'Listen to the sound of time: walking with Saints in an Andalusian village' in Ingold, T. and Lee Vergunst, J. (eds), *Ways of Walking: Ethnography and Practice on Foot* (Aldershot: Ashgate), 93–104.

Macfarlane, R. (2007), 'Afterglow, or Sebald the walker', in *Waterlog: Journeys Around an Exhibition* (Film and Video Umbrella, Arts Council England), 78–83.

Mackenzie, A.F.D. (2006), '"Against the tide": placing visual art in the Highlands and Islands, Scotland' *Social and Cultural Geography* 7(6): 965–85.

Macpherson, H.M. (2008), '"I don't know why they call it the Lake District they might as well call it the rock district!" The workings of humour and laughter in research with members of visually impaired walking groups', *Environment and Planning D: Society and Space* 26(6): 1080–95.

Marples, M. (1959), *Shank's Pony: a Study of Walking* (London: Readers' Book Club).

Matless, D. (1998), *Landscape and Englishness* (London: Reaktion).

Matless, D. (1999), 'The uses of cartographic literacy: mapping, survey and citizenship in twentieth-century Britain', in Cosgrove, D. (ed.), *Mappings* (London: Reaktion).

Mels, T. (ed.) (2004), *Reanimating Places: A Geography of Rhythms* (Aldershot: Ashgate).

Michael, M. (2000), 'These boots are made for walking: mundane technology, the body and human-environment relations' *Body and Society* 6(3–4): 107–26.

Morris, N. and Cant. S. (2006), 'Engaging with place: artists, site-specificity and the Hebden Bridge Sculpture Trail' *Social and Cultural Geography* 7(6): 863–88.

Olwig, K. (2008), 'Performing on the landscape versus doing landscape: perambulatory practice, sight and the sense of belonging' in Ingold, T. and Lee Vergunst, J. (eds), *Ways of Walking: Ethnography and Practice on Foot* (Aldershot: Ashgate), 81–92.

Phillips, A. (2005), 'Walking and looking' *Cultural Geographies* 12(4): 507–13.

Pinder, D. (2005), 'Arts of urban exploration', *Cultural Geographies* 12(4): 383–411.

Shehadeh, R. (2007), *Palestinian Walks: Notes on a Vanishing Landscape* (London: Profile Books).

Solnit, R. (2000), *Wanderlust: a History of Walking* (London: Penguin).

Solnit, R. (2003), *Motion Studies: Time, Space and Eadweard Muybridge* (London: Bloomsbury).

Stacey, R. (2000), 'Facing west: Hamish Fulton in southern Alberta', in Fulton, H., *Magpie: Two River Walks* (Lethbridge: Southern Alberta Art Gallery).

Stevenson, R.L. (1879), *Travels with a Donkey in the Cevennes: an Inland Voyage* (www.bibliobazaar.com/opensource).

Taplin, K. (1984), *The English Path* (Ipswich: Boydell Press).

Tuck-Po, L. (2008), 'Before a step too far: walking with Batek hunter-gatherers in the forests of Bahung, Malaysia' in Ingold, T. and Lee Vergunst, J. (eds), *Ways of Walking: Ethnography and Practice on Foot* (Aldershot: Ashgate), 21–34.

Wallace, A.D. (1993), *Walking, Literature and English Culture* (Oxford: Oxford University Press).

Whitehead, S. (2006), *Walking to Work* (Abercych: A Shoeless Publication).

Wolff, M. (1973), 'Notes on the behaviour of pedestrians', in Birenbaum, A. and Sagarin, E. (eds), *People and Places: the Sociology of the Familiar* (New York: Praeger).

Wylie, J. (2002), 'An essay on ascending Glastonbury Tor' *Geoforum*, 32, 441–55.

Wylie, J. (2005), 'A single day's walking: narrating self and landscape on the south-west coast path' *Transactions of the Institute of British Geographers*, 30, 234–47.

Wylie, J. (2007), *Landscape* (London: Routledge).

Chapter 3
Running: Running as Working

John Bale

Joyce Carol Oates wrote: 'Running: If there's any happier activity, more exhilarating, more nourishing to the imagination, I can't think what it might be' (quoted in Burfoot 2000, 66). On the other hand, a former long-distance runner recalled that by the age of nineteen she had 'already given too much, all my blood and my driving, pounding heart and guts, I cannot possibly keep doing it, giving it more and more again' (Heywood 1998, 97). So, far from supplying the free, joyful idyll of bodily movement invoked by Oates's lyrical representation, the serious runner can be read as a product of ideological, technological and carceral acts that produce a kind of 'human motor' (Rabinbach 1992). The twenty-first century world of running as sport sustains the notion of 'work' and hence contests Anson Rabinbach's suggestion that the twentieth century witnessed the disappearance of the human motor. This chapter, therefore, focuses on running in its utilitarian, rationalised and commodified form. The kind of running dealt with here is one of several similar 'structured mobilities', abstracted and mechanised (Cresswell 2006, 9) and largely focuses on Britain and the West.

Running and racing represent respectively what Yi-Fu Tuan (1998, xii) called 'a ladder of aspiration, at one end of which are the exuberantly or crassly playful and at the other the deeply serious and real', exemplified by the two quotations above. Over time running has aided hunting, helped carriers of messages, provided sensuous pleasure, and been a form of re-creation, fun and entertainment. It has also been a means of punishment and become an occupation within the realm of serious sports. So there are numerous motivations for running.

It is claimed that in the US nearly 38 million people 'run for exercise' (Anon 2008) but my concern is with serious competitive runners, about a million of whom live in Britain (Foster 2004). 'Serious running' is an all-embracing category. Like dancing, there are many classes; for example, sprinting, middle-distance, long-distance, cross-country, hurdling, steeplechasing and road running. During the twentieth century many forms of running were excised from the overall repertoire, exemplifying the rationalisation of this particular movement culture. For example, running backwards is no longer a 'recognised' event, being 'impure' or lacking seriousness. Competitions involving racing on stilts, races between fat men and runners carrying a jockey, and paper-chasing have also been excised (Shearman 1888), examples of 'superfluous motion' (Cresswell 2006, 20), and, I might add, superfluous time and space.

Compared with walking, running seems to have carried negative connotations of disorganisation, chaos, power, danger and damage, and a lack of dignity. Dashing about in an unrefined, boisterous and carefree way contrasts with the romanticised world of gentle walking, signifying calmness and security: Young men running in the cities of nineteenth century Europe would frequently be labelled as mad. More recently, the French philosopher Alain Finkelkraut, mocked the president, Nicolas Sarkozy for his dedication to jogging, noting that 'western civilisation, in the best sense, was born on the promenade' (Bremner 2007, 31), but now often 'relegated to treadmills in climate-controlled *work*out spaces' (Adams 2001, 187). Jogging and running can also be seen as *management* of the body and 'the [runner] says I am in control. It has nothing to do with meditation' (Bremner 2007, 31). Another commentator, *pace* such neo-Marxist perspectives, noted that running is 'about performance and individualism, values that are traditionally ascribed to the Right' (ibid.; see also Brohm 1974; Rigauer 1981). However, running can also be read as resistance and the growth of street and road running has, from the 1980s, contested the dominance of the motorised vehicle and provided a way of claiming back the street for human locomotion. Additionally, an alternative form of running as resistance is the fictive notion of winning by losing, that is deliberately losing a race in order 'to outwit, or win out against winning' (Connor 2005; Sillitoe 1994 [1961]).

Serious competitive running, that forms the basic theme of this chapter, has been recorded from at least Greek and Roman times and in numerous non-European cultures that include the Americas and Africa. The modernisation of running took place at the very time that modernising tendencies were occurring in almost everything from locomotion to music (Kern 1983). These transformations of running can be reflected in its changing landscape and in the bodies of those who run over it. As with other forms of movement-culture, institutions, standardisation and bureaucracies were imposed to control and record a particular quality of running based on rule-bound competition. Among early developments to 'improve' running was the construction of the first purpose-built running track in England in 1837, the foundation of Oxford University Athletic Club in 1866, the Amateur Athletic Association established in 1870, the first modern Olympic Games held in 1896, and the formation of the International Amateur Athletic Federation (IAAF) in 1912, a period in which 'a series of sweeping changes in technology and culture created distinctive new modes of thinking about and experiencing time and space' (Kern 1983, 1).

Having covered some introductory ground I now deal with the concept of 'speed', a central theme in the cultural history of running. Secondly I stress the technologisation associated with the speed-up in running mobility and the question of the body's ownership. Thirdly I acknowledge the inevitability that with the speeding up of running there is also a slowing down.

Running and Speed

Achievement running is work-like and the space of work is essentially directed (Tuan 1974). A project – work or a race – has a beginning and an end. Its logic is characterised by the spatial metaphor, 'linear'. A race is like making a car; it is a project – it starts here (now) and ends there (then). The space is historical and directed. In racing 'What is important is speed – speed in directed space' (ibid., 227). Many observers have read speed as the essence of serious running and in the Olympic triad, *citius-altius-fortius*, the time factor is given priority. Since the late nineteenth century a significant objective of racing has been to achieve ever-faster times with the history of running being read as a 'one way street'; since the nineteenth century nothing has happened to change the fundamental direction of the 'prevalence of speed' (Eichberg 1990, 129). In a race 'an effort is made either to reduce the amount of time that is used, or to increase the amount of space that is traversed' (Weiss 1969, 107). However, attempts to establish the longest distance that can be run in a given time have declined in importance. The record for how far can be covered in one hour, for example, is still recognised by the IAAF but is rarely contested. In an age of instant gratification, most emphasis in track meets is placed on sprint events and fewer 10,000 metres track races take place today than a century ago, suggesting that 'speed and its stresses seem to interest most spectators more basically than endurance' (ibid., 101). The 100 metres sprint can be read as the paradigm of mechanical running, a race that requires total concentration and perfect timing. In long distance running an athlete could fall over, pick herself up and still win but there is no place for mistakes in the 100 metres.

Speed has different qualities. The physiologist A.V. Hill observed in the 1920s that the best kind of speed over a given distance was evenly-paced, suggesting that efficient and rational running required what F.W. Taylor had called 'the organisation of exhaustion' (quoted in Solnit 2003, 212). Another qualitative assessment of running is style. Compared with say dancing (and even walking), style is of relative insignificance during a race, though it is has been far from ignored in manuals of running practice. Concentrating on style at the expense of speed would contradict the objective of maximising performance but it has long been considered acceptable to modify one's style if it improved results. A good style has long been lauded but no prizes are given for style as in gymnastics, diving and dancing. A sports-worker may experience the sensation of speed during a race but victory, a particular performance, or the defeat of a competitor is of greater importance than sensory experiences. The athlete, in training and competition, 'is occupied with the impersonal and the distant' (quoted in Solnit 2003, 212) in contrast to the romantic walker who seeks to carefully observe nature and its wonders. This is not to deny that running in the forest, meadow or beach can encourage the sensation of speed, but in a world where meaning is the sole property of the runner herself, such running is often a means to an end – the competition, the race or the record.

Drawing on Tuan (1977), front space is primarily visual; it is perceived as the future. It is sacred space, towards the horizon, yet to be reached. Backspace is in the past, the profane. Consider the relevance of frontal space to running. Humankind has constantly sought the horizon and the world beyond it and in achievement sports (unlike in play) athletes are thought to constantly seek the record through the generation of greater speed. The record is about conquering distance and compressing space by time. But once the record is found, like the horizon, a new one appears. The experts, who said, as late as the 1930s, that the four-minute mile was impossible, were proved wrong. They had ideas of a fixed limit rather than an ever-receding horizon. As early as 1908 it was stated that a mile could be run in less than four minutes (Dyer and Dwyer 1984). It was not achieved until 1954. Today a time of four minutes is considered a modest performance, having been achieved by high school students and men beyond their fortieth year. In 1900 the fastest time recorded for the mile was 4 minutes 12.75 seconds but as I write (in 2008) it is 3 minutes 43.13 seconds. It seems necessary for athletes to be inculcated with a sense of 'constant lack, of recurring incompletion', Juha Heikkala (1993, 22) arguing that a 'limit cannot be set, because this would collapse the basic structural principle of high-performance sport, which is the constant enhancement of performance'.

Acceleration, Technologies, the Running Body and the Running Landscape

In large part increased speed has been achieved by work-like training methods. Physical work requires the physical organisation of space and, one might add, the body. Particularly associated with running are 'mechanisms' or 'technologies' of various kinds, for example, prosthetics, medicaments and pharmacological supplements such as steroids, the synthetic running track, starting blocks, athletic footwear and special clothing. However, these aids to performance and mobility can be subsumed under the broad concept of training.

Training has been defined as 'that process of education of the body which prepares it to meet with safety exceptional demands upon its energies' when the body 'is called upon to produce a degree of physical output which the circumstances of everyday life cannot supply' (Abrahams and Abrahams 1928, 1). But training was not seen to be indiscriminate. Crucial to success was specialisation. This implied commitment and hard work included a clear schedule for each day of the week. With training came the trainer and the coach – 'experts' who would advise on matters ranging from diet to distance to be run. Coaching takes away some of the runner's responsibility and the coach often possesses a degree of power over his charges. He is also an agent of supervision and surveillance, monitoring and recording his pupils' progress. According to Shearman (1888), training meant 'diet'. In the mid-nineteenth-century two pints of beer a day had been part of the recommended liquid input for a runner in training. During the first half of the twentieth century strychnine and amphetamine were commonly used and by the mid-1960s the world of running experienced the 'anabolic steroid stage' (Spitzer 2004). This was followed by blood doping and the 'post anabolic' stage involving the administration of cerebral and peptide hormones and the (ab)use of natural and synthetic hormones. Associated with doping has been the growing sophistication of training based on physiology and medicine. Research, while initially intended for traditional medical purposes, became applied to improvements in running performance and results (Hoberman 1992).

Various prosthetics have fuelled the 'need' for greater speed and results in modern running. Novel forms of shoes with improved traction were one response to the demand for increased running speed. Nineteenth century competition shoes were made from leather fitted tightly to the foot but because they were not waterproofed the leather stretched making them useless for running. Spiked running shoes were developed by 1852. By 1894 the Spalding Company catalogue featured spiked footwear that were low cut and made from kangaroo leather uppers, the soles having six spikes (Kippen 2007). Progressive changes during the following century reduced the weight of the shoes with their production being dominated by iconic companies such as Adidas, Puma and Nike. Like the bicycle, the running shoe 'alters the body with a faster pair of legs' – not a man and a shoe but 'a faster man' (Kern 1983,113).

Running shoes, like other footwear, are not simply prosthetics or mechanical technologies and such shoes can signify style and identity, embodiments of standardisation, commodification and objectification (Michael 2000) – and the possible exploitation of those who produce them (Schoenberger 1998, 7). After winning an important race, the sprinter Carl Lewis commented that 'a college kid wearing Nikes had beaten the mighty Adidas athletes' – implying that he represented the shoe company (Lewis and Marks 1990, 43). Standardisation is shown by the IAAF (2007a) rules that declare that in shoes worn in races on synthetic tracks the 'part of each spike which projects from the sole or the heel shall not exceed 9mm', reflecting the standardisation of the mundane technology that is applied in the world of serious running.

The stopwatch was another nineteenth century innovation that became a part of the runner's repertoire of technologies. Developed in the 1820s and 1830s the stopwatch sought to 'stop time'. The accuracy of timing improved with the move from manual to electronic timing during the 1930s. The stopwatch as prosthetic was typified by the great Finnish athlete, Paavo Nurmi, who often ran with a stopwatch in his hand (Jukkola 1932). In one of his few allusions to sports Walter Benjamin compared such running 'to the industrial science of Taylorism that employed the stopwatch to analyse minutely the bodily actions of workers for the purpose of setting norms for worker production' (Benjamin in Buck-Morss 1997, 326). The stopwatch turned the competition into a test: 'Nothing is more typical of the test in its modern form as measuring the human being against an apparatus [...] For these reasons the "Olympics are reactionary"'(ibid.). Benjamin noted that 'fascism displayed the physical body as a kind of armor against fragmentation, and also against pain. The armoured [hard], mechanised body with its galvanised surface and metallic, sharp-angled face provides the illusion of invulnerability' (Buck-Morss 1992, 38). It is a body numbed against feeling.

A human form of the stopwatch is the pacemaker, known to *aficionados* as a 'hare' or 'rabbit' (Rosenberg 2005). Pacemaking in races beyond 400 metres (races up to 400 metres are run in 'lanes') is inevitable, simply because somebody has to set the pace. However, contrived pacemaking in which an arrangement is made for one runner to set a pre-determined pace for another runner suggests connivance or a stunt. In chronobiological terms a pacemaker is 'an entity controlling or influencing rhythmic activity' (quoted in Parkes and Thrift 1980, 20) and this model of pacemaking is widely practiced in serious racing. It was used in 1954 when Roger Bannister, in a contrived race with two pacemakers, became the first man to run a mile in less than four minutes. Montague Shearman (1888, 213) deplored such a practice, already widespread in nineteenth century England. His view was that the sooner 'athletes learn that time is a test of speed but nothing else, the better for the sport. The race is not always to the fastest and to possess speed without pluck or judgement is to have very little title to genuine merit'. He would have accused Bannister of 'cowardice' (Shearman 1888, 213). However, the assistance provided by pacemakers 'is a given today with few moral qualms' (Rosenberg 2005).

Starting blocks, 'foot props' or 'bracers' were a further invention that aided sprinters for whom they provided additional impetus and a faster start than small holes dug in the track. Starting blocks were introduced in 1929. Described as 'mechanical aids' they were banned following a world 100 yards record set by the first runner to use them in open competition. While it was recognised that increased speed was inevitable it should not be achieved with the help of 'self starters and superchargers which will act like a shot of dope shoved into the veins of a racecourse' (Perry 1929, 1). They were simply unfair and, after all, the main object of the sport's bureaucracy was 'to *standardize the conditions* in all track and field events' (ibid.). Starting blocks were eventually recognised as a legitimate aid to speeding-up in 1937.

The work-like, industrial and mechanical metaphors that describe serious running may have been most severely critiqued by the neo-Marxist Bero Rigauer who represents competitive running *as* work, particularly in the training regimen to which runners commit themselves. He notes:

> [Forms of athletic training] with their formal division into quantified units of repetitions, distances, times and pauses, demonstrate a direct connection with the industrial speed-up combined with the promise of higher pay [or output or results]. Both systems of behavior can be subsumed under the concept of assembly line production. (Rigauer 1981, 35)

The imbrication of science and running is further illustrated by the application of the photo-scientific work of Marey and Muybridge who sought to identify the 'ideal' running form from their photographic experiments (Solnit 2003, 182).

Tracks and Lanes

In 1868, the English quarter-mile championship was contested at the Beaufort House grounds in London. Edward Colbeck won the race:

> Coming along at a great pace, he led all the way round the ground, and was winning easily when a wandering sheep found its way upon the path and stopped still there, being presumably amazed by the remarkable performance which the runner was accomplishing. The athlete cannoned against the sheep, broke its leg, and then went on and finished his quarter in 50 2/5 seconds. (Shearman 1888, 87)

This incident illustrated the need for territorialisation – the separation of runners' space from that of animals and spectators. Additionally, the layout of the track itself required considerable modification. To avoid runners clashing with each other visible lines were inscribed on the track to keep athletes in their place for races of less than 800 metres. A variety of bold white lines that might appear meaningless to anyone but a runner or fan find meaning for the athlete. They are lines of demarcation. They are, for the duration of some races, personal spaces. Serious runners are subject to strict rules. Step over these lines and you are disqualified from the race. As Marshall Berman (1983, 117) quotes the official corrector of books in St Petersberg: 'geometry has appeared, / land surveying encompasses everything. / nothing on earth lies beyond measurement'.

During the nineteenth and much of the twentieth centuries track surfaces were grass, clay, dirt or cinder. However, these were subject to deterioration during severe weather conditions, despite meticulous attention by groundsmen and the applications of 'turf science'. From the mid-1960s running tracks have increasingly been made of synthetic materials that make them more productive as machines – that is, they are efficient and rational, 'producing' more running and faster records and, like the street, can be read as part of the 'process of *appropriation* of the topographical system' (de Certeau 1988, 97).

The global bureaucratisation of the running track represents the erosion of difference and its replacement by a space constructed with global specifications that come close to 'placelessness' (Relph 1976). Such a condition was recognised as early as 1857 at the first Cambridge University sports. The quarter mile course was circular and inscribed on a cricket ground. It was reckoned that there 'was no scenery to admire and no time to admire it if there had been' (Abrahams 1956, 159). This approximated to the ideal model for foot racing: 'Ideally a normal set of conditions for a race is one in which there are no turns, no wind, no interference, no interval between starting signal and start. No irregularities in the track – in short, no deviations from a standard situation' (Weiss 1969, 105). To this list could be added standardised dimensions of the running track itself, domed stadiums and the tendency to eliminate 'nature'.

Until the last quarter of the twentieth century the spatial dimensions of the running track varied considerably. During the nineteenth century many tracks had lengthy straightaways so that speed could be maximised but tracks with tight bends constrained speed and encouraged the possibility of injury. As a result, gentle bends and shorter straight sections become the standard model. However, the size of the track varied considerably, some having a circumference of 300 metres while others were 600 metres. Standardisation was required as large tracks were thought to improve speed and provide an unfair advantage for those who sought records. Consequently, the quarter-mile or 400 metres circuit became the maximum size for the recognition of record results. On oval-shaped tracks 'starting and terminating points are clearly marked, but in racing the destination itself has no inherent significance; it can indeed be identical to the starting point' (Tuan 1974, 237) and it is 'as though the ideas of time and competition become more and more real, the actualities of place and substance less and less so' (Solnit 2003, 182).

Improved performance and results have been attributed, in large part, to the synthetic running track that became an established part of sports-running mobility in the 1960s, pioneered by a subsidiary company of the giant American chemical firm Monsanto. Analogous to the tarmacadam road surface and later the motorway that served to speed-up the mobility of automobiles, the synthetic track can be read as another form of prosthetic. Such additions to the human body – ranging from shoes to tracks – reveal an 'interconnection between the human body and the wider world that signals the arrival of the prosthetic subject' (Cresswell 2006, 167) – in this case the prosthetic runner-athlete.

Serious modernised running 'requires exactly specified and formalized environments, for in most cases the dominance of territory or mastery of distance' (Wagner 1981, 92). Consider the exactitude and precision found in a couple of the many rules for the construction of running tracks, as defined by the global arbiter, the International Association of Athletic Federations (IAAF 2007a). For example, a record shall be made on a track, the radius of the outside lane of which does not exceed 50 metres, except where the bend is formed with two different radii, in which case the longer of the two arcs should not account for more than 60 degrees of the 180 degrees turn. Additionally, because synthetic surfaces lose thickness in service, by wear and weathering, it is recommended that, upon completion of the track, the average thickness of the synthetic surface should be at least 12 mm. Nowhere will the surface be less than 10mm and the total area with a thickness between 10 mm and 10.5 mm shall not exceed 5 per cent of the total surface area. Such globally imposed rules infer that the running site should be the same, exactly the same, as any other in the world. Generally regarded as the founding father of the modern Olympics, Baron Pierre de Coubertin was well aware of the implications of such technologies. He noted that the ancient runners ran in sand in order to increase the level of difficulty and, hence, increase their status. Modern runners, on the other hand, seek to make the race easier, in order to increase their speed. He also noted that, for example, spiked running shoes took the performance and result away from the athlete. Writing in 1936 Coubertin continued,

> suppose we could imagine shoes, or even tracks, with springs that would somehow throw the runner forward with each step. In this case it is not just the movement that is being made easier, but some of the athletic effort would be done by the equipment the athlete is using. The speed achieved in this way will not be entirely his own. (Coubertin 2000, 200)

Concerns about the cyborg runner surfaced in 2007 when the legitimacy of a runner with prosthetic limbs was questioned due to the possibility that his artificial limbs might give him an unfair advantage over 'able-bodied' runners. South African runner Oscar Pistorius was born without a fibula in both legs, which were amputated between his ankle and knee. His prosthetics are made up of sockets fitting over the stumps of his legs and a stiff carbon-fibre blade bolted to the outside of the sockets (Powell 2007, 97). The governing body for track and field (IAAF) was forced to decide how much of an athlete's speed was the result of one's own training and initiative and how much of it was the result of someone else's technology. Ultimately, such a discourse relates to the on-going question about who *owns* the (runner's) body and whom the athlete represents.

However, the assumption that such technological 'developments' result in time-space compression can be viewed as a form of technological determinism (Stein 2001, 107). Certainly, the consideration of *only* technological improvements ignores the social processes in which technology is embedded. For example, improvements in running records (i.e. faster times) may have resulted from the fact that more people have been taking up running over time or that diet has improved. Concomitantly, it seems plausible that if fewer people took part, or dietary standards declined or an anti-sport regime banned achievement-oriented sports, time-space compression might be replaced by time-space expansion (i.e. slowing down).

It would be an illusion, however, to believe that technological 'progress' in attempts to speed up athletic mobility is reaching its limits; indeed, genetic engineering and other experiments on athletes' bodies have barely begun (Lundberg 1958). There are various ways of encouraging even greater running speed through geometric and architectural changes in facilities such as tracks and arenas (Dyer and Dwyer 1984), making them even more efficient as, *pace* Le Corbusier, machines for running.

Toward Slowness and a Change of Scale

The 'track' and the 'lane' no longer connote the rustic. The modifications of body and landscape have been in the interests of increased speed. Yet there are several types of slowness that can be recognised in running. For example, the slowing down with age in a sporting career or technical slowness with athletes spending most of their training by moving slower than they do in competition. Speed can be reduced by technology, slowing it down via the slow-motion image and the photo-finish camera, while the stopwatch seeks to stop clock-time altogether. Furthermore, strategic slowness occurs when the greater part of a competition is performed slowly for tactical reasons. Slowness can also be seen as constrained time, i.e. sports' rules. A race tells us, not the speed that is possible, but the speed that is possible under 'antecedently defined conditions and commonly accepted rules' (Weiss 1969, 111). Finally, slowness can be seen as a form of time-space expansion.

Mobility and speed are said to be central to running because they are central to the record. Writing in the late nineteenth century, however, Shearman (1888, 205) was upset by the record mania of his day and was 'heretical enough to believe that the worshipping of records is idolatrous, and inconsistent with the creed of the true sportsman'. Coubertin would have disagreed and subscribed to the *citius, altius, fortius* ethic. However, he recognised the tension between *citius* and the inevitability of excess and much serious running can be read as 'sanctioned excess'. Some of Shearman's thoughts resonate with running at the present time. He deplored records that were made with the help of pacemakers, or a top-class athlete who competed against modest opponents. Races, he felt, should be between athletes of equal ability with the emphasis placed on performance rather than result. 'The sooner, therefore, that athletes learn that time is a test of speed but nothing else, the better for the sport. The race is not always to the speediest, and to possess speed without pluck or judgment is to have very little title to genuine merit' (Shearman 1888, 205).

The de-emphasis on records and speed sounds strange in the twenty-first century but it is not without support. For example, Sigmund Loland (2000) argues for greater moderation and more subjective, qualitative progress in running, arguing that running has become too specialised and tests an increasingly narrow range of skills. He argues, therefore, for less specialisation, modelled on Grand Prix type meetings where races could be held over different types of courses – beaches, forests, track, uphill etc., hence insisting on a broader range of physical skills.

The *slowness* that Loland implies in his advocacy of moderation has, perhaps, failed to gain the attention that it deserves. In a world frequently dominated by speed, slowness, even in an athletic victory, is not generally respected. Despite the positive reactions to slowness noted above, it is often seen negatively and athletes have, on occasions been reprimanded and in some cases punished for slow performances. Two examples will suffice to show how slowness can be seen as a deplorable performance. In 1954 a Soviet middle distance runner, Vladimir Okorokov, competing in the national 1,500 metres championship, recorded a time of 3 minutes 49.8 seconds in his heat but (only) 3 minutes 54.6 in winning the final. His relative slowness (an effective strategy to achieve victory) resulted in him being denied his national title and his prize. An official pronouncement stated: 'Instead of running to the best of his possibilities, he let others set the pace and just forged ahead in the last few yards to win a cheap victory' (*Athletics World* 1954, 93). And a respect for slowness was also absent at the Goodwill Games of 2001, held in Brisbane, Australia. At this event, two 5,000 metres races were held, one for women and another for men, a common practice in the highly gendered character of top-class running. The women's race was won in a faster time (15 minutes 12.22 seconds) than the men's (15 minutes 26.10 seconds), an outcome that was generally deplored. The reporter for *The Guardian* newspaper went so far as to declare that the slowness of the men's race 'went beyond farce and brought the sport into disrepute'. The crowd issued 'boos and catcalls' and the athletes ran 'at little better than a jog with no pacemaker to stir things up' (Mackay 2001).

These stories, and the responses to them, reveal the significance of both speed and the result in achievement running. Performances, tactical though they might have been, and with exciting sprint finishes, are felt to be devalued if they are characterised by overall slowness. To race is not enough. For athletes to run slowly is to transgress the norms of modern running. However, despite the negativity implied by slow running, it should be recognised that acceleration is a relative concept; there cannot be a speeding-up without also a slowing-down (May and Thrift 2001). Roger Bannister's 3 minutes 59.4 seconds for the mile made previous running, by athletes such as Nurmi, look much slower than it had previously appeared.

A geographically sensitive approach recognises that running in certain places (nations) shows clear signs of slowing down and that national patterns of record-breaking vary considerably. There is no speeding-up without an associated slowing-down. It appears that slowness has replaced acceleration in athletic performance in some nations where athletes are running slower than previously and some national records have stopped being improved upon. Put another way, in some places the perfect race seems to have been run. For example, at the time of writing the British 800 metres men's record has not been improved upon since 1981 and the 5,000 metres since 1982. From the 800 metres to the marathon, British runners' results are standing still or slowing down (Gillon 2007) and today's performances are profane in comparison with the apparently sacred records of the likes of Sebastian Coe. There are many cases of nations where such records have stood still (Butler 2007), exemplifying what Edward Ullman (1974) called 'dead time'. To be sure, the record is often an exceptional result while average times may be speeding-up, though this is not always the case. Even so, it is the record to which spectators crave and to which athletes aspire though it is often argued that the value of records is over-emphasised (Weiss 1969, 166).

Such decline is not the result of technological factors. It appears from the western European and North American perspective that the kinds of running that are generally declining in speed are those that require hard, manual labour, i.e. work. In real terms there has been a decline in the standards of British elite and club runners since the 1980s. It has been suggested that a major cause of slowness is that:

distance running – an activity which involves endurance, struggle and pain – rests at odds with the current popular discourse of the attenuated self: the vulnerable individual who needs to be protected by an overbearing state. The second cause is that running is being choked by charity raising and the anti-competitive ethos that goes with it. (Buckingham 2005, np)

A more sedentary western society that recognises the exhaustion and hard work involved in modern sports and the dietary habits of the western world (the 'obesity crisis') has something to do with the slowing down described above. Rabinbach's suggestion that the 'human motor' has disappeared seems to have some credence in the slowing down that I have noted above and a drift away from hard, work-like running. However, while European and North American runners are slowing down those of East Africa are speeding up. In 1923 Coubertin (2000, 702) argued that 'the time has come for sport to advance to the conquest of Africa […] and to bring to its people the *ordered* and *disciplined* muscular effort, with all the benefits which flow from it' (italics added). With the subsequent help of the IAAF, this mission has proved to be remarkably successful and a large number of postcolonial nations now participate in globally organised running. East African runners now dominate the most work-like or 'industrial' forms of running. For example, in 2006 twenty per cent of the world's hundred fastest 1,500 metres runners came from Kenya alone, similar percentages characterising other distance events (IAAF 2007b). This East African presence reflects a form of mobility that assumes global dimensions.

Conclusion

Anson Rabinbach (1992, 296) concluded *The Human Motor* by noting that numerous critics have defined postmodern society in terms of 'the sense of a declining interest in work [and] increasing leisure' and that with 'the declining significance of industrial work as a paradigm of human activity and modernity, the body no longer represents the triumph of an order of productivism' (ibid., 300). However, modern running, with its emphasis on human labour and the production of records and results, makes the running body closer to, rather than further from, the world of work (Shilling 2005, 101). This suggests that the 'human motor' is running smoothly for those sports-workers who have internalised the industrial ideologies of this particular form of mobility.

References

Abrahams, A. (1956), *The Human Machine* (Harmondsworth: Penguin).
Abrahams, H.A. and Abrahams, A. (1928), *Training for Athletes* (London: Bell).

Adams, P. (2001), 'Peripatetic imagery and peripatetic sense of place', in P. Adams et al. (eds), *Textures of Place: Exploring Humanist Geographies* (Minneapolis: University of Minnesota Press), 186–207.

Anon (2008), 'Study reports size in number of runners', http//fitnessbusinesspro. com/news/rise_runner_numbers/ (accessed 20 August 2008).

Athletics World (1954), 2(11): 93.

Berman, M. (1983), *All that is Solid Melts into Air* (London: Verso).

Bremner, C. (2007), 'More Rimbaud and less Rambo, critics tell sweaty jogger Sarkozy', *The Times*, 4 July.

Brohm, J.-M. (1974), *Sport: A Prison of Measured Time* (London: Ink Links).

Buckingham, A. (2005), 'Running isn't just for fun', *Spiked*, http//www.spiked-online.com/index.php?/site/article/1133/ (accessed 13 September 2007).

Buck-Morss, S. (1992), 'Aesthetics and Anaesthetics: Walter Benjamin's artwork essay reconsidered', *October*, 62, 3–41.

Buck-Morss, S. (1997), *The Dialectics of Seeing: Walter Benjamin and the Arcades Project* (Cambridge: MIT Press).

Burfoot, A. (2000), *The Runner's Guide to the Meaning of Life* (New York: Rodale).

Butler, M. (2007), *Athletics Statistics*, www.athleterecods.net/ (accessed 18 July 2007).

Connor, S. (2005), 'My fortieth year had come and gone and I still throwing the javelin', *Static*, 1, http:static.londonconsortium.com/issue01/connor_beckett. html (accessed 14 November 2005).

Coubertin, P. de (2000), *Olympism: Selected Writings* (ed. N. Müller) (Lausanne: International Olympic Committee).

Cresswell, T. (2006), *On the Move* (New York: Routledge).

de Certeau, M. (1984), *The Practice of Everyday Life* (University of California Press: Berkeley).

Dyer, K.F. and Dwyer, T. (1984), *Running out of Time* (Kensington, NSW: University of New South Wales Press).

Eichberg, H. (1990), 'Forward race and the laughter of pygmies', in Mikulás Teich and Roy Porter (eds), *Fin de Siècle and its Legacy* (Cambridge: Cambridge University Press), 115–31.

Foster, A. (2004), *Moving On* (London: Sports Council/UK Athletics).

Gillon, D. (2007), 'Distance decline analysed', *Athletics Weekly*, 7 May, 26–7.

Heikkala, J. (1993), 'Discipline and excel: Techniques of the self and body in the logic of competing', *Sociology of Sport Journal*, 10(4): 397–412.

Heywood, L. (1998), *Pretty Good for a Girl* (New York: The Free Press).

Hoberman, J. (1992), *Mortal Engines* (New York: The Free Press).

IAAF (2007a), *Competition Rules, 2006–2004* (Monaco: IAAF), http://www.iaaf. org/newsfile/23484.pdf (accessed 3 July 2007).

IAAF (2007b), http://www.iaaf.org/ (accessed 13 September 2007).

Jukkola, M. (1932), *Athletics in Finland* (Helsinki: Söderström).

Kern, S. (1983), *The Culture of Time and Space 1880–1918* (Cambridge: Harvard University Press).

Kippen, C. (2007),*The History of Sports Shoes*, http://podiatry.curtin.edu.au/sport. html (accessed 3 July, 2007).

Lewis, C. with Marks, J. (1990), *Inside Track* (New York: Simon and Schuster).

Loland, S. (2000), 'The logic of progress and the art of moderation in competitive sports', in T. Tännsjö and C. Tamburrini (eds), *Values in Sport* (London: Routledge), 39–56.

Lundberg, K. (1958), *The Olympic Hope* (trans. E. Hansen and W. Luscombe) (London: Stanley Paul).

Mackay, D. (2001), *The Guardian Unlimited*, September 7, http://www.guardian. co.uk/o,6961,,oo.html (accessed 10 January 2002).

May, J. and Thrift, N. (eds) (2001), *Timespace* (London: Routledge).

Michael, M. (2000), 'These Boots are Made for Walking: Mundane Technology, the Body and Human-Environment Relations', *Body and Society*, 6(3–4): 107–26.

Parkes, D. and Thrift, N. (1980) *Times, Spaces and Places* (Chichester: Wiley).

Perry, L. (1929), *Glendale News Press*, June 12. http://frankwycoff.com/aau_ disputes_foot_props.htm (accessed 6 September 2007).

Powell, D. (2007), 'Blade runner prepares to take on the best', *The Times*, 29 June, 97.

Rabinbach, A. (1992), *The Human Motor* (Berkeley: University of California Press).

Relph, E. (1976), *Place and Placelessness* (London: Pion).

Rigauer, B. (1981), *Sport and Work* (trans. A. Guttmann) (New York: Columbia University Press).

Rosenberg, D. (2005), '"Rabbits" in Running Races: Pseudo-Competitors in Pseudo-Competitions', paper presented at 33rd Annual Meeting of the International Association for the Philosophy of Sport, Palacky University, Czech Republic.

Schoenberger, E. (1998), 'Discourse and practice in human geography', *Progress in Human Geography*, 22(1): 1–14.

Shearman, M. (1888), *Athletics and Football* (London: Longmans).

Shilling, C. (2005), *The Body in Technology and Society* (London: Sage).

Sillitoe, A. (1994 [1961]), *The Loneliness of the Long Distance Runner* (London: Flamingo).

Solnit, R. (2003), *Motion Studies* (London: Bloomsbury).

Spitzer, G. (2004), 'A Leninist monster: Compulsory doping and public policy in the GDR and the lessons for today', in J. Hoberman and V. Møller (eds), *Doping and Public Policy* (Odense: University Press of Southern Denmark), 133–45.

Stein, J. (2001), 'Reflections on time, time-space compression and technology in the nineteenth century', in May and Thrift, *Timespace*, 106–20.

Tuan, Y.-F. (1974), 'Space and place: humanistic perspective', *Progress in Geography*, 6, 211–52.

Tuan, Y.-F. (1977), *Space and Place* (Minneapolis: University of Minnesota Press).

Tuan, Y.-F. (1998), *Escapism* (Baltimore: Johns Hopkins University Press).

Ullman, Edward (1974), 'Space and/or time: Opportunities for substitution and prediction', *Transactions of the Institute of British Geographers*, 63, 125–39.

Wagner, P. (1981), 'Sport: Culture and Geography', in A. Pred (ed.), *Space and Time in Geography* (Lund: Gleerup).

Weiss, P. (1969), *Sport: A Philosophic Enquiry* (Carbondale: University of Southern Illinois Press).

Chapter 4

Dancing: The Secret Slowness of the Fast

J.D. Dewsbury

'Dance is what suspends time within space'.

(Badiou 2005, 62)

A black void. A sense of presence. Light slowly presents a figure in the centre of the stage. Behind him seven cream slats face the audience and 'wall' the back framing the stage space for us. The slats are suspended from the ceiling but hover stable and fixed seven metres from the front of the stage giving room for something to happen 'between'. Dance is precisely the room for the body of the dancer 'to be'. Presently the slats are equally lit and touch one another – there is no gap between them although you sense there could be. Four stage lights are preset and visible at the front apron of the stage, two on each side – one of these directs its light directly towards the slat across from it, the other directs its light at an angle, hitting a slat further across closer to the middle of the stage. Presently these front lights are off. The dancer moves in slow and deliberate tai chi like manoeuvres: he is alone and cuts a melancholic figure: searching, trying to find his feet, stunned but murmuring still. Presently, in co-ordination with a choreography of emerging light coming from the lights on the front, the dancer moves towards the front left hand side. The two lights on that side slowly illuminate and project light towards the slats at the back: they also of course project the shadow of the dancer. We now have three dancers on stage – the flesh of the dancer at the front left of the stage and two of his silhouettes cast and captured against the surface of the slats at the back. The different angle of the lights means that three different dances unfold each with intimate echoes of the other although they are located differently from within the same spatial movements. Powerfully, the solitude that the dancer initially presented dissipates as he communes with his shadows on the 'slatted' wall behind. He dances with more purpose and spirit. However, as he moves to approach his shadows, one seems to run away from contact whilst the other merges with him as he closes in to embrace it.

Across the disciplines of the social sciences and humanities, the practice and craft, the beauty and therapy of dance has increasingly been cited as an exemplary site for research experience (Cancienne and Snowber 2003; Cooper-Albright 2003; McCormack 2002, 2003), conceptual debate (Thrift 2000; Gil 2002, 2006; Badiou 2005; Counsell 2006), and as political and cultural expressions for questioning our embodiment (Martin 1992; Foster 1995; Nash 2000; Revill 2004; Cresswell 2006). What distinguishes dance here is its quality as a disciplined practice often apprehended through two different approaches: first, through a concern for the *historical context* of the trained bodies produced, the spaces in which these are facilitated and through which such bodies are constituted; and second, through an emphasis on the *performative show* of the delicate precision and complexity – subtle, elusive and tantalizing – of the body itself in movement, which is at once both communicative of both its vitality and passivity, its suffering and enjoyment. To think of dance as the art of mobility par excellence signals that the art of dance rests in its contrast between the quick and the still, the fast and the slow, the performative and the historical, the non-representational and the representational. In this it presents 'the between' itself, where 'the between' gives a sense of the performance vitality pivoting off the ground that is its stage. To an extent you cannot have one without the other, a performance without a stage, but it is easy to concentrate on one at the expense of the other and miss the critical performative *movement* on show. Presently, we are seemingly more required to and acquired in disposing ourselves quite deliberately within productions emerging out of an increasingly explicit engineering of the relationship between our bodies and a whole set of cultural markers, technological enhancements, social narratives and commodified promises. Set against this age of modernity dance affords us space to think more precisely about the ontology of movement. The cultural, technological, social and capitalist relationships overflowing the body – be they medical, pedagogical, psychological, and quite simply everyday – all measure and mete out a disposition of and, for me, away from the body. So the uniqueness of dance is that it is *also* an apprehension and manifestation of what the body is outside of these relational flows: as Derek McCormack puts it in his work on dance therapy, this 'also' draws attention to the 'connective sensibilities' of dance that are 'procesually enactive, as styles and modes of performative moving and relating rather than as sets of codified rules' (2003, 489). Dance thus presents the vital presencing of the flesh's heartbeat and what we 'see' in our neural vibrations, blockages, nerve snaps and internal image screens, when confronted by dance is also scaled at the level of the quick without code. Dance is that which becomes coded but is not itself code in its point of incessant emergence. Dancers are coded, genres are coded, dance choreography is coded, but the *dance itself*, like all art forms when considered in their pure state, manifests the power of the present to disrupt, disclose, and expose these codifications. Herein, whilst the body may be the marker and the container of these codes of representation with all their politics the body is not itself representational. It is because of the body's presentational status that dance is so difficult, perhaps precisely impossible, to textually represent and interpret. This is also why dance attracts public and academic alike in its atmosphere of vital seduction: that which is seemingly available to know and to interpret always has further enticements eluding knowing. Such elusory movement is another form of, and space for, politics.

What I am primarily aiming at in this chapter is to focus attention on the metaphysical and philosophical *show* of dance. As such I want to turn to dance here as a spectacle, as a performance event experienced by an audience, and I am not going to dwell on its history, its therapeutic capacity, or its cultural reception. Rather for me, more importantly:

> dance is an appearance, if you like, an apparition. It springs from what the dancers do, yet it is something else. In watching a dance, you do not see what is physically before you – people running around or twisting their bodies; what you see is a display of interacting forces, by which the dance seems to be lifted, driven, drawn, closed or attenuated, whether it be solo or choric ... The forces we seem to perceive most directly and convincingly are created for our perception; and they exist only for it ... Anything that exists only for perception, and plays no ordinary, passive part in nature as common objects do, is a virtual entity. It is not unreal; where it confronts you, you really perceive it, you don't dream or imagine that you do. (Langer 1951, 341–2)

The importance here lies in the promotion of what I see to be the non-representational aspect of dance: namely its show not of the physically actual but of the virtually perceived. There has been a decade long steady engagement with dance within geography now, and as my approach is towards the non-representational (see Thrift 2008), I want to situate some of my concerns within this broader engagement. Nigel Thrift turned to dance in its capacity 'as a contemplation which values improvisation and encourages attunement to emergent form' (Thrift 2000, 237). Pushing on from dance's power to instil this contemplative reflectivity Thrift focused more on dance as a specific expression and practice of thought-as-action which offered 'a symptomatology of movement which can help us to both understand and create expressive potential by gesturing to new ground' (Thrift 2000, 238). Dance is thus seen as 'a technique for creating new forms of awareness and persistence' through the 'nonmimetic and non-representational' which can then subsequently become an alternative 'kind of identity citation' (Thrift 2000, 240). At the same time, but not in critical tandem, Catherine Nash pertinently critiqued Thrift's appropriation of dance. Nash began by arguing that:

> While Thrift's project is clearly to use dance to develop a wider set of ideas around body-practices, performativity and 'nonrepresentational theory', dance theorists have already raised a series of issues that may temper as well as inspire the growing interest in the geography of embodiment, performativity and play. (Nash 2000, 656)

I am wary of attempts to temper things: whilst it means to qualify and modify, it is also suggests a sense of bringing into 'a better state' the points of debate. Herein, for me the slip between body-practices and embodiment is telling: embodiment speaks of the ways in which the body itself is translated and framed by social formations whereas body-practices are more to do with what the body actually does. So whilst there is important work done on embodiment we need also to investigate body affects themselves, and I do not think we need to operate towards a critical synthesis here. That said, it is dangerous to separate the body out from the ideologies that make it mean something, not least because it implies some originary or essentially pure body. I do think however that something is missed if you don't pause on the difference between the terms. Thus I agree with Nash that dance should not be seen as 'a new version of an old division between thought and action, between mind and body' (Nash 2000, 657; an argument echoed in Revill 2004). But I don't agree with the fast move to suggest that an attempt to think dance in terms other than 'the deeply social character of coded performances of identity' signifies 'asocial implications' pertaining in 'noncognitive embodied practice' (ibid.). I accept the responsibility as an academic to uphold an awareness of the danger that non-representational approaches to dance can lose 'the sense of the ways in which different material bodies are expected to do gender, class, race or ethnicity differently' (Nash 2000, 657). But I will counterpoint the suggestion that 'this sense of dance beyond language appears unable, despite the stress on relational selves, adequately to combine a sense of the social or of social relations with the unanalysable world of the precognitive or prereflective' (ibid.). The understanding of the social here is too sure footed, and I believe that we need to question more what brings sociality about. As Paul Harrison puts it 'there is a dimension of corporeal existence which is not reducible to what we as social scientists call the social, at least no more than it is reducible to some natural drive or tendency' (2008, 441–2). Therefore I do not think then that dance is 'always mediated by words as it is taught, scripted, performed and watched' or that it is primarily 'often highly formalized and stylized'; and although of course 'even untrained dance is culturally learnt and culturally located' (Nash 2000, 657) my simple point is: *what* is being culturally located? And *how* is the cultural here being learnt? I think dance shows us this 'what' and 'how' in ways that are much more innocent than Nash thinks possible. Jose Gil's work reflects this 'innocence' when he argues that the work of Merce Cunningham rejects expressive conventions and an advocation of the autonomy of music and movement as it works towards 'the principle that one can render movement in itself, without external references' (Gil 2002, 117; Cunningham 1992).

In what follows, the chapter is structured in seven short sections before a conclusion. The seven sections echo the seven stages of humankind (echoed from Maliphant's seven slats in *Shift*) and indicate that I take dance's show to be a universal one. Each marks out the sense that accretions are made upon the body that take us ever further away from any sense of there being an innocent, unscarred, uneducated, pure body available for mobility despite the belief that that is what is at stake in dance's show. Pointedly such accretions are non-representational as much as they are representational. Therefore the performance of mobility here, exemplified as dance, is marked to show the interferences between thought and action, from first sensations of live flesh (*baby*) to the abstracted pain where bodies fast learn from experience (*toddler*), the ever present capacity to carve out territories of the imagination in movement innocent to the actual world itself (*child*), the propulsions of movement in desire that are as much metaphysically constitutive as physically compromising (*teen*), the training infrastructures that channel, augment and denigrate capacities to act (*young adult*), the affirmative fact that worlds disappear as unpredictably and as consistently as they appear (*adult*), and that where there is a body there is a virtual universe of possible affective extension to this world (*elder*).

1. *Baby* – Live Flesh

What is pledged on the stage of dance? The scream-breath: the first breath we take as we are removed from our mother's womb. Dance presents the sheer effort of life and as such is staged upon the vulnerable flesh, which is fragile, susceptible, pulsing within, and dripping without at birth, of blood. Subsequently this flesh as body fast becomes nascently self-aware of its existence. This self-awareness comprises itself of sensations such as pain and pleasure with flesh as body acting as the site of their immediate cause. As yet, however, the baby's body is a disorganized body, disorganized to the material connections that surround it – be they of code, the organic, or reflexive thought (however basic, at this stage of life, that sensibility to move is). Dance on one level presents this very vulnerability of birth, of our incessant birth to presence, as a not yet ordered body, or rather as the always alternative potential of the body *in situ* regardless of the current framing set up of what a body may do right here, right now. Conceptually, Deleuze and Guattari's 'body without organs' as a connective field of flesh and nerve captures the sense of the production of movement of this pared down body. Thus:

> a wave flows through it and traces levels upon it; a sensation is produced when the wave encounters the forces acting on the body, an 'affective athleticism', a scream-breath. When sensation is linked to the body in this way, it ceases to be representative and becomes real; and *cruelty* will be linked less and less to the representation of something horrible, and will become nothing other than the action of forces upon the body, or sensation (the opposite of sensational). (Deleuze 2003, 45)

2. *Toddler* – Abstracting Pains

It would be wrong to suggest though that dance absolutely presents to us the body itself full of potential with equally essential or available capacity and skill. Tiredness, stomach pains, nausea, headaches, breathlessness are part of the body's fabric (see Bissell 2009). Equally, we soon realize whenever we enter a new room, grow up, approach again a second time, that we are inevitably being registered differently from the last time, based on the previous room, experience, *that* last time branded now somehow with some form of identification. Finally, then, we are subjectified, lost or at home, careful and restricted or carefree and sanguine, welcomed or rejected. But, dance can be understood as a more naked form of presentation if we think of the nakededness of being denuded of the codes of strata that so define us and place us in our social role with all the politics of culture that that entails. As already implied above, I acknowledge Deleuze and Guattari's three key codifications or strata: the organism, 'signifiance' (or interpretation), and subjectification.

> You will be organized, you will be organism, you will articulate your body – otherwise you are just depraved. You will be signifier and signified, interpreter and interpreted – otherwise you are just deviant. You will be subject, nailed down as one, a subject of enunciation recoiled into a subject of the statement – otherwise you are just a tramp. (1988, 159)

In terms of this denuded articulation dance manifestly stutters a sense of mobility and meaning in its expression of verticality and attraction (Badiou 2005, 58). Dance movements can be seen to echo, underline, or be based upon those first steps, those first mobilities towards basic need: food and companionship, defecation and rest. As animals we slowly emerge and frame ourselves by code: at one end of the spectrum dance shows the perfection of such codings in movement (ballet) whilst at the other (more abstract forms of modern dance) it dwells on those basic movements that both literally and empathetically 'move' us. Dance then also enacts an abstraction from traditionally understood codifications; it abstracts us from the socially presented body as potentially depraved, from the identifiable body as the always also deviant, and from the located body that belongs as the incessantly nomadic tramp.

What these abstractions suggest is that dance is an art skilled in a heightened minimalism which thus presents the pared down simplicity involved in learning to move for the first time. More expansively and fundamentally, consider that dance is produced through the body only then needing ground and space to work. Dance has little need for other materiality: no requirement of paint and canvas, film and light, text and prop, clay and moulding tool, paper and ink. And, whilst dance is often associated strongly with music – the argument being that the jurisdiction of music's rhythm and beat is a tangible subjection that produces the dance – it can also be argued, as Badiou does, that 'when it comes to dance, the only business of music is to mark silence' (Badiou 2005, 62). In other words, music is less the agency of the ink and more the paper on which the dance is written. Again, a minimal hauntology of what is not physically present but rather virtually perceived: dance is about not yet knowing our way but having a sense of the direction we should be taking. We are always working off stratifications (we wouldn't need them otherwise), but stratifications are not fixed preset channellings of movement.

3. *Child* – Innocent Territories

If we are without such fixities and absolutely preset co-ordinations we are forever entering innocent territories of association:

> (The child) is innocence and forgetfulness, a new beginning, a sport, a self-propelling wheel, a first motion, a sacred Yes. (Nietzsche 1969, 55)

I want to take on the question of innocence first before considering the productive operation, or logic, of forgetfulness. 'Children never stop talking about what they are doing or trying to do' (Deleuze 1998, 61) – a verbal manifestation of that groping hand in the dark trying to find the light switch. In other words, a disorientation, an innocence of location, a lack of self-awareness to others or other meanings other than the immediate exploration of movement in place or milieu (darkness, space, a vague knowledge that there is a switch to give light) in which one finds oneself. It is then a trajectory making that 'merges not only with the subjectivity of those who travel through a milieu, but also with the subjectivity of the milieu itself' (ibid.). What I want to suggest here is that dance is the subjectivity of the milieu itself. Innocence is then precisely a lack of concern with subjectified identity, an indifference to the self in terms of any meaning given to form that self from outside the taking place of immediate movement. The mobility that is mapped here then is not an extensive one. Dance is not showing mobility 'in relation to a space constituted by trajectories' so much as the intensity and density of movement 'concerned with what fills space, what subtends the trajectory' (Deleuze 1998, 64). Dance is made out of its affects, both active and passive, and is rendered as a staging of movement that is about its affective constellation, not its co-ordinated meaning. As an art dance 'attains this celestial state that no longer retains anything of the personal or rational … (it) says what children say' (Deleuze 1998, 65).

The final paragraph in Deleuze's short essay 'What children say' is the tour de force of the piece and it frames the logic of forgetfulness in dance:

> To an archaeology-art, which penetrates the millennia in order to reach the immemorial, is opposed a cartography-art built on "things of forgetting and places of passage". (Deleuze 1998, 66)

This place of passage is the alternative situation of mobility that I am arguing is made manifest in dance. First, it has no other memory than that of the material of which it is made: thinking of dance, the material is that of the organic body. Second, there is no origin of this material to be remembered; rather the assignment of this place of passage is to make the displacement of the material something visible – i.e. in dance to show the body disappearing. This is precisely the effort of any performance art, that whilst its ontology is akin to our corporeality, that it is live and inevitably disappears (see Phelan 1993), it is that making appear that is so impressive, even if, or precisely because, it only lasts for a split second. Third, it marks the fact that the performance 'does ... not exist before the work' (Deleuze 1998, 67). Dance therefore shows the emergent and ephemeral, the effective and active, territory of movement performance itself. Take this oft quoted Deleuzian example of a child's territory of reassurance in the movement of a song:

> A child in the dark, gripped with fear, comforts himself by singing under his breath. He walks and halts to his song. Lost, he takes shelter, or orients himself with his little song as best he can. The song is like a rough sketch of a calming and stabilizing, calm and stable, centre in the heart of chaos. (Deleuze and Guattari 1988, 311)

This singing becomes a refrain, a repeated melody or gesture that marks out a territory, that calmness in the fear, that meaning and orientation where there is none as such. Quite exactly, a virtual realm, real but not actual: a sensation existing only for perception there and then. 'The refrain may assume other functions, amorous, professional or social, liturgical or cosmic; it always carries the earth with it' (ibid., 312). This basic composition of space and ground situating the body as the live flesh of our viscerally bare being-there, wherein our materiality itself *feels* as opposed to knows itself in the world. As such the refrain – and this connects the child's song to the dance – is 'a melodic formula that seeks recognition and remains the bedrock or ground of polyphony' (ibid.) for whoever enters into, or witnesses as dance spectator, its affective territories of rhythm.

4. *Teen* – Exposing Desire

> Dance is innocence, because it is a body before the body. It is forgetting, because it is a body that forgets its fetters, its weight. It is a new beginning, because the dancing gesture must always be something like the invention of its own beginning. And it is also play, of course, because dance frees the body from all social mimicry, from all gravity and conformity. A wheel that turns itself: This could provide a very elegant definition for dance. Dance is like a circle in space, but a circle that is its own principle, a circle that is not drawn from the outside, but rather draws itself. Dance is the prime mover: Every gesture and every line of dance must present itself not as a consequence, but as the very source of mobility. And finally, dance is simple affirmation, because it makes the negative body – the shameful body – radiantly absent. (Badiou, 2005, 58)

The affirmation of dance comes with the idea that the dancer's body radiates an absence of shame; an absence of a precondition judgement. Effectively, in its abstraction as a figure, the dancer's body has shaved off traces of shame in becoming the body of desire. This desire takes mobility to be being dis-posed. Dance shows us dispositions then, the ever emergent mobility without consequence. It is also an ontological statement that even when still, seemingly motionless, we are still nevertheless dis-posed – there is internal movement because it is the ontological condition of the body as such to be that which is always in-disposed. Conceptually, the work of Jean-Luc Nancy explicates this sense of disposition, for his

> argument draws us not to an interiority or an enclosed space, in which we turn away from the world and back into ourselves, but opens us toward a disposition, in the sense of a mood, an inclination, an attunement or *Stimmung*, in which we acknowledge neither the origin of communication nor some transcendent source that legitimates all communication and governs its rules, but a "disposition of Being". (Armstrong 2009, 118)

I think we can legitimately take dance to be the presentation of this disposition of Being 'where it is no longer a matter of presenting being (to oneself), but of being toward being-as-act, of touching on the emergence – or being touched by the coming – of being-as-act' (Nancy 1997, 51). Dance is then not quite just a beautiful mesmerizing staging of disappearance it is more precisely 'at the junction (conjunction? collision?) of the given and the desired' (ibid.) – a never fulfilled moment. It is scripted not by lack but by loss, a loss that is constitutive of the moment, of being touched, of being disposed, of being outside oneself, of being a being of ek-static existence. Whilst dance presents the ek-static nature of existence, to argue this does not represent a naïve rendering of dance as an exhibition of 'a primitive ecstasy' or the 'forgetful pulsation' of the body – dance is quite precisely opposed to this 'spontaneous vulgarity of the body' (Badiou 2005, 60). That is not to say that some of the best modern dance I have seen ignores drawing attention to the hot erogenous zones of folded flesh and orifice that signal the potential intensification of lines of flight of energy, fluid, touch, scent and gaze. Dance touches on these limits, and even when argued for in non-representational terms, it does not cross these limits into a realm of 'unreflected, unarticulated, practical action' (Revill 2004, 201); rather it shows them in their limitation. As bodies, dance shows us that in our bodies we are constantly being disposed, always at the limit of what is conceivable, and as such these limits are not erased in dance:

> You have to keep enough of the organism for it to reform each dawn; and you have to keep small supplies of significance and subjection, if only to turn them against their own systems when the circumstances demand it, when things, persons, even situations, force you to; and you have to keep small rations of subjectivity in sufficient quantity to enable you to respond to the dominant reality. Mimic the strata. You don't reach the BwO and its plane of consistency, by wildly destratifying. (Deleuze and Guattari 1988, 160)

5. *Young Adult* – Training Figurations

A critical junction arrests a difference of emphasis between representational and non-representational approaches to embodied activities, and aptly dance is cited as exemplary for both sides of the argumentative divide. Tim Cresswell's research on dance plays up the importance of 'movement in dance specific normative geographies' to argue that it might be better to value 'thinking of mobilities as produced within social, cultural and, most importantly, geographical contexts' (Cresswell 2006, 56). We cannot escape the fact that institutional frames are everywhere, scaffolds for straightening our backs. Even in the context of minimalist modern freeform dance the spatializing material structure of the stage instructs or affords specific movement potentials (after Deleuze 2003, 20). I then have no problem with the argument that the apparent lightness and effortlessness presented in the show of dance hides the fact that dance is technically and painfully acquired and honed as a skill, and often through and for different ideological means. When wowed by dance we secretly know the effort behind the staging of a never quite so perfect body: an important reflection to remember in the commodification of the body beautiful and dynamic. Dance has an occluded background: it intensively depends on rehearsal offstage for the preparedness of choreographic codification, and it extensively depends on rigorous training to produce the precisely intelligent body of the dancer. Starkly, we must not forget that there is a pastness to the presence of performance that excludes the audience: there is suffering behind the bliss. But the show of dance is all that most of the audience see.[1] So I still feel it valid to push for an appreciation of the event of dance on show – this is what the training is for, namely to be able to exhibit a heightened experience by giving room through body movement alone for some virtual expression to become manifest, by improvising into emergence new forms of association and sense communication, and finally herein exposing us to the limits of what the body, not everybody, *can* do.

6. *Adult* – Event: Undecided Disappearance

As skills gather and preparations synthesize towards a consistency for a performance, a cultural product which is the choreographed 'show' comes to presence. The argument here though is not to forget what the live event of dance actually shows: namely, at least, the visceral movement of attraction/repulsion, the

---

1   I say most as of course trained dancers in the audience will wince through the experiential traces in their own body resonating with the moves on stage that their flesh knows the cost behind; and likewise they will be more attuned to the different styles of choreographer, country and body type behind the drafting and redrafting of the specific productions and moves that so affect; and non-dancer audience members will slowly educate themselves in their company.

rhythmic flow of movement that takes place 'without-thinking', and the carving out of different territories in rhythms and spaces between and outside of bodies. More prosaically, whilst it is commonplace to sense the importance of the space between bodies it is rarely articulated as such:

> We all understand something about communication and movement, and the placement of human beings in a space. That sounds really abstract but it isn't. If you go into a restaurant or a bar and it is empty and you choose where to sit, you are making a very clear decision. If someone comes in and they sit too close to you and you are the only people in the room that says something. That is about an uncomfortable body language, about placing yourself too close to somebody in a particular situation. (Anderson, *Double Take DVD*, 2005).

The show of dance is the articulation of this spacing. As adults we are delineated by spatially written rules inscribed in the flesh as a kind of felt knowing (culturally differentially learnt of course). These rules determine a repertoire of action but do not condition it. They are like an affective field where:

> the field condition that is common to every variation is unformalized but not unorganized. It is minimally organized as a polarization. ... The field of play is an in-between of charged movement. It is more fundamentally a field of potential than a substantial thing, or object. (Massumi 2002, 72)

The field potential for a dance event is the marked out territory of the dance space. These territories are materialized in dance via empathetic connections with which a new 'commune' of association and understanding is possible. One of the key qualities of dance is the ability to create radically different worlds just by changing the dancers' gait or gestures, and in so doing completely dissolving their personalities, their egos and their human qualities in one twist of a torso or in the merging of contrapuntal rhythms. This affective field, or event-potential of a world produced by a different communicative mode, can be exemplified through the dance work of choreographer Lea Anderson. Anderson choreographs dances for her all female ('The Cholmondeleys') and all male ('The Featherstonehaughs') dancers, sometimes combining both troupes. In celebrating twenty years of her work she took dances choreographed for the men and got the women to dance them in a retrospective show called *Double Take* (see Jennings 2004). One of the dances was 'Elvis Legs' where one of the dancers announces at the start, to demystify the process of dance more as a music band performer would, 'we spent hours and hours in the studio with thousands of Elvis videos and we watched them over and over again, and then we picked out our favourite arm and leg movements, and put them back together, the wrong way' ('The Cholmonderleys', *Double Take* DVD).

Three female dancers walk onto the stage dressed in elite city suits tailored for women. The stage is marked out by pink tape on the black floor probably eight metres square with a red cabaretesque curtain behind. They stand in a diagonal, the closest (Anna pictured below) front right, and proceed to dance a series of Elvis signature dance gestures although they never resemble nor mimic directly the King himself. Anna leads the movements initially, these are echoed and refrained by the two dancers behind, sometimes they are in sync creating a powerful affective verve, other times different gestures re-emerge in contrapuntal movements. Coming in and out of sync suggests shared territories, as do the echoes of a gesture from an early movement danced by the dancer at the back as Anna the dancer at the front proceeds with putting the movements together again in a faster rhythm.

As Anderson herself puts it: 'People get a bit of an idea of where it has come from without worrying about what it means, so they can just experience the movement without thinking that there is some secret to it that they don't understand, that they have to be enlightened, in order to enjoy' (*Double Take DVD,* 2005). For me it was precisely an affect of enjoyment: noticeably Anderson just says 'to enjoy' not enjoy 'it'. That sums it up precisely for me: here was an event of dance that produced a zone of affective connection that was 'to enjoy', there was nothing in itself to enjoy. There was rather just an experience of affirmative attraction.

> Movement is neither a displacement nor a transformation, but a course that traverses and sustains the eternal uniqueness of an affirmation. Consequently, dance designates the capacity of bodily impulse not so much to be projected onto a space outside itself, but rather to be caught up in an affirmative attraction *that restrains it.* (Badiou 2005, 59)

Badiou proceeds to argue from this that the 'essence of dance is virtual, rather than actual movement' (2005, 61). Much like the dynamic image presented by Langer back in the 1950s, Badiou points to the idea that dance, as it shows quickness and lightness, exhibits more precisely slowness and restraint. This means for him that what it gestures in movement points to a 'strange equivalence' with nongesture such that dance indicates 'that, even though movement has taken place, this taking place is indistinguishable from a virtual nonplace' and that whilst dance 's composed of gestures that, haunted by their own restraint, remain in some sense undecided' (Badiou, ibid.). *The gestures may have the spirit of Elvis but what they show, and what territories of affect they now produce, is something else.* Again, this finds resonance in Gil's argument that 'because there is an imbrication or overlap between the played representation and the referent, dance always preserves a non-representational element that escapes the production of signs' (2002, 126). And subsequently this pushes home the role of the metaphysical power of dance, that:

> Dance would provide the metaphor for the fact that every genuine thought depends upon an event. An event is precisely what remains undecided between

the taking place and the non-place – in the guise of an emergence that is indiscernible from its own disappearance. (Badiou 2005, 61)

7. *Elder* – Affective Extensions

Throughout the chapter I have been wrestling with the line of argument that you cannot separate the dancer from the dance. Maybe I am wrong but the counter-argument persuades me still: there exists the dance itself. Again take the notion that whilst the body expresses emotion in dance, it also expresses itself. So I want to state quite strongly that it is the dance itself that affects us most because as Badiou puts it 'dance visibly transmits the Idea of thought as an immanent intensification' (2005, 59). Or if read through Deleuze, dance presents the elevation of mechanical forces to sensible intuition (see Deleuze 2003, 46). This is argued across dance studies where the foundation of our embodiment, culturally inflected for sure, suggests we house the ability, proprioceptively, to connect with another person's bodily feelings communicating in terms of a kinaesthetic empathy (see Foster 1995; Makula 2006). This is not an essentialism as it is not an absolute condition but rather an indication that to a greater or lesser extent we share some abilities to sense. Elsewhere this is what Martin termed communicating through 'sentience' (1992) rather than through sign; as a consequence the parameters of interpretation, meaning and the efficacy of dance is altered. Thus:

> While repeatable, it [the dancing body] represents nothing in and of itself. It is not a signifier to any signified. Rather, the movement communicates through its kinetic effects. Kinetic effects, the stimulation of the senses or sentience, are feelings expressed directly from one body to another and amongst a group of bodies. Hence, to study the experience of dance is to isolate both the unique communicative aspects of the body and the moment of pure action of an unrepresented (and unrepresentable) subject. (Martin 1992, 10)

In other words, then, we can render mobility in aesthetic form as that which presents communication through kinetic effects. In a central chapter in his book on dance, *Movimento Total: o corpo e a danca,* Gil presents what he calls the 'paradoxical body' of dance. This pivots off the idea that 'the dancer does not move in space, rather, the dancer secretes, creates space with his movement' wherein 'a new space emerges' which is phrased quite plainly as 'the space of the body' (2006, 21). This space is folded and constituted by what we can notionally refer to as objective space but crucially it transfigures it as 'a scene invested with affects and new forces' (ibid., 22). Thus dance's show is not just restricted to aesthetic concerns as we can see when Gil extends his argument:

The space of the body is not only produced by gymnasts or artists who use their bodies. It is a general reality, present everywhere, born the moment there is an affective investment by the body. It is akin to the notion of 'territory' in ethology. As a matter of fact it is the first natural prosthesis of the body: the body gives itself new extensions in space. (ibid.)

Conclusion – Still Points

It is not so much the historical context or cultural reference that denies dance the status of being the ephemeral art per se, rather it is from within the question of immanence upon which the 'showing' of dance gathers its felt meaning, that provides the live presence of dance with so many resonances suggestive of already learned knowing. From this still non-representational perspective several arguments arresting traditional and academically habitual ways of considering the body, performance and event have been made for proceeding towards this presence of dance itself. First, dance is the art form least reliant on representation as it primarily works with and demands space or spacing alone. Second, the dancing body is never someone as in a Hamlet, nor even as Russell Maliphant, because the dance itself 'is born under our very eyes as body' (Badiou 2005, 64). Herein the body-subject dissipates as alterations in movement emerge not through a centralized will but through the affect of space, rhythm and relation with other bodies (see Gil 2002).

Third, dance shows the exposure of the body in its disclosure, in how it discloses us to the world. 'Dance, as a metaphor for thought, presents thought to us as *devoid of relation to anything other than itself*, in the nudity of its emergence. Dance is a thinking without relation, the thinking that relates nothing' (Badiou 2005, 66). This is to emphasise that the show of dance, outside of its event, may precisely put nothing into relation – that is for me part of its power to set you thinking.

Finally, and this is the place where I was always at with dance, as a spectator: 'dance does not address itself to the singularity of a desire whose time, besides, it has yet to constitute' (ibid.). There is no object of dance for it is always in a state of disappearing. Dance is then 'the permanent showing of an event in its flight, caught in the undecided equivalence between its being and its nothingness. Only the flash of the gaze is appropriate here, and not its fulfilled attention' (Badiou 2005, 67–8). The argument parallels that of Nancy, where what is thought in this citing of dance is a singular-plural spacing of sense as open and disclosing, as that which makes intelligible in its emergence but cannot be finally seized, delimited nor fully understood. Thus the presentation of dance (or the 'coming to presence of any being') 'is itself a (singular) plural event, a spacing or sharing of the multiple passage of sense'; it is 'a spatial-temporal event of sense' (James 2008, 108). If dance suspends time within space, this suspension is precisely this spatial-temporal event of sense – in terms of mobility *per se*, dance shows the secret slowness of the fast, all those many territories and interferences that expose us between actions. Thus, what dance does represent is up to the impact it has on and through your own particular singular body, but that body is clothed in an immanence of plural shared sensation – that which you 'see' which is not physically before you – be that affect, Deleuzian infinitives (to smile, to love, to cry, to hurt) or the syntactical grammars of so many everyday relations that are genealogically and experientially wired into the body over time (gestures, proximities, pauses, excitements).

I sat there stilled and astounded, profoundly moved by eight minutes of dance. I had let go of the tears that had welled up in my eyes, having given in to my failed attempts to blink them back, a failure cushioned, and perhaps more readily indulged, by the dark gloom of the auditorium, and they streamed down the sides of my face: I was overwhelmed, outside of myself, not quite in control: it felt beautiful: I was sadly happy: I felt alive to my life for, as insignificant as it is, I saw there something of it on stage. Admittedly life had made some present cuts into me that precisely resonated with what had been on show: the returning presence of my long dead father as those shadows dancing on the stage, echoed and brought to the fore in the metaphorical recent death of another, losing the love of the other whom I loved. So I too felt alone on stage but somehow searching for a presence which was always with me (my body stands like my dad, apparently). Surrounded by infinite possibilities we are all equally finite. That is what dance showed me that night.[2]

References

Armstrong, P. (2009), *Jean-Luc Nancy and the Networks of the Political* (Minneapolis: University of Minnesota Press).

Badiou, A. (2005), *The Handbook of Inaesthetics* (Stanford: Stanford University Press).

Badiou, A. (2007), 'A theatre of operations: a discussion between Alain Badiou and Elie During', in *A Theater Without Theatre* (Museu d'Art Contemporani de Barcelona), 22–7.

Bissell, D. (2009), 'Obdurate pains, transient intensities: affect and the chronically pained body', *Environment and Planning A*, 41, 911–28.

Cancienne, M.B. and Snowber, C.N. (2003), 'Writing rhythm: movement as method', *Qualitative Inquiry*, 9(2): 237–53.

Cooper-Albright, A. (2003), 'Matters of tact: writing history from the inside out' *Dance Research Journal*, 35/36(2/1): 10–26

Counsell, C. (2006), 'The kinesics of infinity: Laban, geometry and the metaphysics of dancing space', *Dance Research*, 24, 105–16.

---

2   The dance show this experience refers to was: *Shift* (1996) – Choreographer and Dancer: Russell Maliphant; Lighting Designer: Michael Hulls; Music: Shirley Thompson.

Cresswell, T. (2006), 'You cannot shake that shimmie here': producing mobility on the dance floor', *Cultural Geographies*, 13, 55–77.

Cunningham, M. (1992), 'Space, time and dance', in Kostelanetz and Anderson, *Merce Cunningham: Dancing in Space and Time* (Pennington: A Capella Books).

Dean, P. (ed.) (2007), *Hunch: Rethinking Representation* (London: Episode Publishers).

Deleuze, G. (1998), *Essays Critical and Clinical* (London: Verso).

Deleuze, G. (2003), *The Logic of Sensation* (London: Athlone Press).

Deleuze, G. and Guattari, F. (1988), *A Thousand Plateaus* (London: Athlone Press).

Foster, S.L. (1995), 'Choreographing history', in S.L. Foster (ed.), *Choreographing History* (Bloomington: Indiana University Press), 3–24.

Gil, J. (2002), 'The dancer's body', in Brian Massumi (ed.), *A Shock to Thought: Expression After Deleuze and Guattari*. (London: Routledge), 117–27.

Gil, J. (2006), 'Paradoxical Body', *The Drama Review* 50(4): 21–35.

Harrison, P. (2008), 'Corporeal remains: vulnerability, proximity, and living on after the end of the world' *Environment and Planning A* 40, 424–45.

Highwater, J. (1992), *Dance: Rituals of Experience* (3rd ed.) (New York: Oxford University Press).

James, I. (2008), *The Fragmentary Demand: an Introduction to the Philosophy of Jean-Luc Nancy* (Stanford: Stanford University Press).

Jennings, R. (2004), 'Whatever way you say it, they can still cut it 20 years on'. [Online] Available at: http://www.guardian.co.uk/stage/2004/may/23/dance [accessed: 16 December 2008].

Langer, S. (1951), 'The dynamic image: some philosophical reflections on dance', in Walter Sorell (ed.) *The Dance has Many Faces* (New York: World Publishing).

Mackerell, J. (2004), 'The Rite Stuff', [Online] Available at: http://www.guardian.co.uk/stage/2004/apr/14/dance [accessed: 16 December 2008].

Makula, P. (2006), 'The dancing body without organs: Deleuze, femininity, and performing research', *Qualitative Inquiry* 12, 3–27.

Martin, R. (1992), 'Dance as a social movement', *Writings on Dance* 8, 9–20.

Massumi, B. (2002), *Parables for the Virtual: Movement, Affect, Sensation* (Durham, NC: Duke University Press).

McCormack, D.P. (2002), 'A paper with an interest in rhythm', *Geoforum*, 33, 469–85.

McCormack, D.P. (2003), 'An event of geographical ethics in spaces of affect', *Transactions of the British Institute of Geographers* 28, 488–507.

Nancy, J.-L. (1997), *The Sense of the World*. (Minneapolis: Minnesota Press).

Nash, C. (2000), 'Performativity in practice: some recent work in cultural geography', *Progress in Human Geography* 24, 653–64.

Nietzsche, F. (1969), *Thus Spoke Zarathustra* (New York: Penguin).

Phelan, P. (1993), *Unmarked* (London: Routledge).

Revill, G. (2004), 'Performing French folk music: dance, authenticity and nonrepresentational theory', *Cultural Geographies* 11, 199–209.

Thrift, N. (2000), 'Afterwords', *Environment and Planning D: Society and Space* 18, 213–55.

Thrift, N. (2008), *Non-Representational Theory: Space, Politics, Affect* (London: Routledge).

Williams, R. (1977), *Marxism and Literature* (Oxford: Oxford University Press).

Films

Double Take – a double bill of 'Flesh & Blood' and 'Double Take' (The Cholmondeleys and The Featherstonehaughs, 2005)

Chapter 5

Driving: Pre-Cognition and Driving

Eric Laurier

Introduction

The most significant recent developments in the study of car travel have come from mobility studies rather than the long dominant field of transport psychology (Cresswell 2006; Hannam et al. 2006; Sheller and Urry 2006). In sociology John Urry (2000, 2004) has argued that we need to rethink how we conceptualise society in tandem with how we understand travel. Just as society is characterised by its increasing mobility as it begins the twenty-first century; so mobility, in the form of transport, is manifest in building and maintaining extended networks of colleagues, friends and family (Urry 2003). In geography Tim Cresswell, over a number of works, has charted how the notion of movement might require a much more fundamental shift in how we investigate spaces and places (Cresswell 2006). In a variety of ways, the study of car transport has found itself re-emergent in this new field. At the same time car travel is, of course, a pressing problem for a world with rapidly rising levels of car ownership and use, unprecedented levels of energy consumption, pollution and road congestion. A pressing problem that cannot easily be solved for the very reason that the car itself is the solution to so many of our daily logistical problems: getting to and from work, shifting groceries, collecting children from school, visiting friends and family and going on holiday (Larsen et al. 2006; Pooley et al. 2006).

In a recent ESRC project called *Habitable Cars*[1] myself, Barry Brown and Hayden Lorimer have examined how we move as groups in our cars in the UK (for other car cultures see (Miller 2001)). Rather than building automobility up as the system that system-theorists would choose, one so global and entrenched that we are quite unable to stop its steady march, our approach is to disperse theory into fields of practical action, while also accepting the frustratingly dispersed and dispersing nature of mobility. Thrift's approach to car travel in 'Driving in the City', has been a tremendous inspiration in investigating driving-in-traffic for at least five reasons:

1. He takes automobility seriously as a *central form* through which an actor-networked version of everyday life has been *re-organised* over the past 100 years. Cities the world over, and not just Los Angeles and Las Vegas,

---

[1] For more details on the project: www.ges.gla.ac.uk/users/~elaurier/habitable_cars/

     are rebuilt around the car, in terms of the architectures of motorways, the garage's relation to the house, the arrival of the non-place petrol stations and more.

2. He is interested in the various ways we have of *living* in the car, understood to be 'profoundly embodied and sensuous experiences'. The inter-twining of car and person, as a novel and irreducible form of what Mike Michael (1998) investigated as 'co-agency', which organises not only how we move but emotional responses (road rage being the obvious example used by Michael and others (Katz 1999; Lupton 2002)).

3. He has dealt with the car itself as a highly designed space which can be investigated in its steadily evolving *fit to* and subtle *transformations* of its driver. Software and ergonomics steadily build new forms of 'humanization' (Thrift 2004a, 10), where Husserlian phenomenology finds itself becoming a building programme for car makers, used to refashion how we sense the weight, speed, sound and touch of the car. GIS, GPS and other forms of computing developments also leading to the possibility that driving may converge with practices of walking in terms of displaying to other vehicles an account of our movements on the road.

4. He touches upon the practices *other* than driving that happen there, which we, in common with Thrift, have called 'passengering' (Laurier et al. 2008). A rich vein of ways of talking, gesturing and looking at one another and the landscape of the road.

5. Unusually for a human geographer, he continues to draw on literature from cognitive theory and neuroscience. Re-drawing consciousness, thought, decision-making and action via theories of the non-cognitive realm that exists in the time before we ruminate upon it post-hoc.

On that last point on the list, the car is a perspicuous setting for the consideration of the boundary between cognition and pre-cognition for two further reasons, the first being that things happen very fast on the road and drivers respond in split seconds. This is not the thoughtful scene of writing where the author has hours to try and form their next paragraph, can delete it without harm and revisit their passage several weeks later and change it all over again. Car crashes brook the most limited revisions. Should a car pull out in front of you, as you would say afterwards, 'I had no time to think about it, I simply had to swerve'. The second reason is that when, for instance, in driving our children home from school we are occupied with so many other tasks, such as stopping the children arguing and planning what to cook for dinner, we 'let the car drive itself' (as one of our project participants put it). Under such circumstances we will find ourselves, on autopilot, taking the left turn to the school when we should be turning to the right that day to go to the swimming pool. Somehow our actions appear to have continued before our mind catches up and notices the error.

    Driving and the automobile system has also served as a touchstone in Thrift's other works (Thrift 1996; Thrift and French 2002). His wide ranging body of work on spatial practice (2000, 2004a, 2005b) has a more longstanding concern with the relationship between a series of entangled pairs: action with knowledge, pre-cognition with cognition, and, the subconscious and anteconscious with consciousness. In each of these pairs, the refiguration of what would be either philosophical, psychological or psychoanalytic binaries are re-arranged through a history of technological change and innovation. For Thrift, how a spatial practice becomes possible at a certain period, and not others, really matters and what new possibilities dawn as new machineries emerge, such as software that intervenes in driving, matters still more.

In seeking to redress the desire of social theorists to treat those they study as proto-social theorists, or worse cultural dopes, and thereby to over-intellectualise ordinary practices, Thrift cites research by cognitive scientists of various stripes on the 'pre-cognitive' or 'non-cognitive'. The importance of the pre-cognitive is laid down in several places in Thrift's work (2000, 2005a, 2005b) and in each of these its use is bound up with embodiment, the senses, affect, nervous systems, a non-Freudian sense of unconscious action, and technological backgrounds, a central one being the transportational background of mobile societies.

When we interact on the road and in our cars there can be two sorts of effects, one being cognitive: knowing, thinking, recollecting and perhaps imagining as joint actions, the second effect being action that pre-figures and in some way anticipates thinking, reflecting, willing and believing. The cognitive domain is one that we might read across to the intellectual and intellectualised practices and concepts that implicate us as investigators of mobility and that we map on to more mundane mobile worlds. A domain that equips us and that we are well equipped to act in. What we seem to be missing is what mobile bodies of various kinds can do before they get to knowing what they are doing, thinking about what they are doing, recollecting what they or others were doing and imagining what they or others might do, are doing, have done.

If the pre-cognitive is movement that is somehow *before* thought, if it is action that precedes any sort of thinking about or deciding to initiate that action, then we find possibilities for new forms of control of movement and new resources for fashioning alternative possibilities. Once we accept that there is such a thing as the pre-cognitive then it can become both a domain of the 'technological unconsciousness' and un-noticed yet active background that soft capitalism is pursuing in terms of channelling it into the functioning of mobile societies (Thrift 2004b) and also a human capacity that is being transformed and extended by, amongst other Latourian things (1996), car software and ergonomics.

To summarise, new kinds of sensing have become possible. Reach and memory are being extended; perceptions which were difficult or impossible to register are becoming routinely available; new kinds of understated intelligence are becoming possible. These developments are probably having most effect in the pre-cognitive domain, leading to the possibility of arguing that what we are seeing is the laying down of a system (or systems) of distributed pre-cognition (Thrift 2005b).

A scintillating account, one that certainly has us wondering whether we are being moved by automobility in its most sophisticated technological forms. Moved to act by these reconfigured gatherings of engines, ergonomic programs and navigational aids before we have a chance to consider deciding whether we want to be moved. Having alerted us to these, by parts, exciting and terrifying expansions of sentience into zones of human movement that appear to be pre-cognitive, Thrift urges us to reconsider our joint-agency with other humans, animals and, of course, machines.

We will return to this critique later in the chapter after examining, in some detail, a transcript of a video clip of 'passengering' and driving gathered and analysed as part of the Habitable Cars project. One purpose being that we then have some worldly material in hand not only to learn about 'passengering' but also to re-specify what 'intentions', 'thinking' and 'action' in driving are (away from driving examples, see Edwards 1997). Secondly we will re-engage practices of reasoning as they relate to the practicalities of driving a car (Livingston 2006). In the section that follows we will be examining two commuters who car share and have been doing so for a couple of years. As we join them, it is morning and they are about five minutes into their journey to work.

'I thought'

P= passenger, D= driver[2]
((Approaching slip road))
((P raises his hand to point at car pulling out))
((His finger then touches nose))
P: I thought he was gonna::, *((looks into car as they pass it))* aye, she was gonnae pull out and go for it

---

2   Denoting these people as these particular categories of actor already begins to make assumptions about the relevant categories, where they may or may not be relevant. I can only signal such concerns here, for a fuller examination of omni-presence or not of categories (see Schegloff 2007; Ochs 1979).

P: Somebody pulled out in front of me
+
((*turns toward driver*))
(1.0)
D: ((*turns toward passenger*))
P: Comin' in
((*both return to looking forwards*))
(1.0)

A first thing I would like us to take from this fragment is that the passenger is monitoring the road ahead in tandem with the driver. The second thing is that the passenger has pointed out something. There are all manner of things that passengers point out, such as cloud formations ahead, new houses being built, unusual vehicles on the road and more. What is special here is that the car being pointed out here is relevant to the driving of the vehicle: a car on a potential collision course. That passengers are involved in monitoring the road, at all, is a feature of the socially organised nature of driving road vehicles that cognitive psychology, with its focus on the lonely brain locked inside its skull with only nerves, muscles and lenses to look out upon the outside world, generally excludes for the purposes of its studies (Potter 2005; Potter and te Molder 2005). Rather than concentrate on demonstrating the frequency or pertinence of passenger involvement in driving cars I would like us to examine this fragment in the light of the moral restraints on how much, and exactly when, a person, encumbered with the rights and responsibilities of a passenger, can intervene in the driver's work.

If you are a passenger, no matter how talented a driver you are, if you intervene too often or inappropriately, you can justifiably be accused of being a 'back seat driver'. As Watson (1999) puts it, any remarks made about driving by the passenger are 'inferentially rich'. The fact that cars are driven by one person rather than two (or four, or a team) is more a matter of histories of car design and social convention than an issue of individualised cognition. Nevertheless because it is so, driving the car has become the exemplary scene of psychology's demonstrations of various models of thinking, action, automaticity and so on (e.g. Groeger 2000) How then might a passenger offer a warning or make an observation about the road that is pertinent to the driving's course and not be overstepping the mark? There are several possibilities: one that echoes Watson's (1999) ethnographic study of truck drivers on un-metalled roads is noticing an object that we can project as collidable with, given our course. Pointing-at other vehicles or objects, even on collision courses, is nevertheless relatively rare because should a passenger point them out they are, at the same time, raising the matter that the driver might not have noticed them. The driver can respond to such remarks as criticism of their competence in driving the car.

At the outset the other car has *not yet* pulled out but, in the finely judged projections of its course, its location and motion in relation to the visible markers of the road is cause for concern. Even afterwards, quite what it was up to is unclear.

However its anticipated course is into the road rather than stopping at the give-way junction. Even though the driver may well be monitoring what the other car is doing, the absence of him, for instance, slowing down or even turning his head slightly toward the other car deprives the passenger of any visible response, and thereby appreciation of the threat posed by the car's rapid approach and over-stepping of the road markings.

Before we therefore assume that the driver has either failed to notice the other car or failed to produce an appreciable response that would have headed off the passenger's intervention, we can consider the further possibility that what we may be party to is a classic game of 'chicken' on the road. A game where the driver is banking on the car seeing him keeping-on-coming, where if he slowed down the other car might exploit that as an opportunity to jump into the road and accelerate away leaving behind the smell of burning rubber and an outraged driver. I am not saying this will happen or am sure that it *is* what's happening but merely saying that one of the ways drivers handle one another is to get the other one to stand down. Moreover drivers exploit the visibility of the absence or presence of noticing one another. Where both cars have rights to go ahead, particularly in seriously congested roads such as those of Mexico City (Sormani 2004), they will try and avoid letting the other driver see that they have seen them. If one has not seen the other, then for the one who has seen the other, they ought to let them through because they cannot rule out that the other driver genuinely has *not* seen them.

What is happening here is taken up by the passenger as that genuine circumstance of the driver not having seen the other car. The car which approaches the junction could quite imaginably not have looked properly to the left or have had this car in their blind spot and missed its approach. Under such circumstances the approaching car should then save the day by stopping in time and thus preventing a potentially fatal crash. It is worth remembering that one of the main reasons for there being actually very few crashes on the road is that where one driver makes an error their error is noticed by approaching drivers and repaired by them. In the majority of situations it takes two drivers' inattentiveness to have a crash.

Let us return to Thrift and pre-cognition via the passenger's gesture: a point that is transformed into a touch of the chin, and then the hand is taken out of sight, what Lerner and Raymond (forthcoming) call a midcourse pivot. It is tempting to equate gesture with what is before or outside of language, an embodied intelligence. The brain is reacting, communication is underway before the brain has a chance to reason about what the body is doing. In a neuro-scientific pre-cognitive account, the passenger's brain is firing up first as it works out what is going on and then sending the electrical charges down his nervous system to get that arm moving, the finger extending to link it with the optical information being processed by the brain from the eyes, and so on, and so psychologically forth.

All of this sounds highly plausible because we are so deeply immersed in cognitive psychology and popular neuroscience that such an account sounds not merely plausible but a matter of scientific fact. To argue that it is not so requires careful retracing of the many ways in which Cartesian rhizomes have run through language to link up with our concepts of thinking, acting, free will and perception. As Wittgensteinian scholars of the mind will remind us, treating 'seeing', 'noticing', 'pointing out' (as a passenger moreover) as brains processing retinal information and activating muscles is a conceptual confusion, a problem with incorrect use of our language (Coulter 1983; Edwards 1997; Hacker 1996; Watson 2003). The 'brain' does not work out the danger the car at the junction poses through processing retinal information and sending messages to its limbs, it is a 'person' that works out the danger the car at the junction poses. If we ask who saw the car, the correct response is the passenger, not the passenger's brain. This is not in any way to deny that their whole body is indeed involved and they could not do what they do in the way that they do it without arms, fingers and a brain. It is to say that if we ask who is pointing out to who, it is the passenger pointing out to the driver, not brain one pointing out to brain two.

Rather than continue to critique Cartesian neuro-scientific accounts here what I will do is make brief remarks about the gesture in its course. To do so is to begin to give us a sense of how we might inquire into gestures as part of the production and reception of reasonable and reasoned courses of action. The pointing happens quickly, where pointing can linger on an object to make sure that there is something to be seen, is seen, and what is being pointed at, is seen. It should be in three sequentially related parts: one being its emergence, the second its pause and the third its dissolution, or perhaps transformation. In three parts it allows for displaying that something is about to be pointed at, then with lingering and thereby picking an object out, and then an evaluation of that thing. As it actually happens, the pointing never pauses, the hand is made into the shape, the finger rises up but doesn't stop, like a bus rushing past a bus-stop, Because the pointing runs on it thereby re-appraises what has been picked out. We make sense of the almost-pointing not as a semiotic gesture (as-it-were) where the hand is making a sign for the car, we make sense of it by locating what it could be locating in the visual field ahead of the car (Goodwin 2003).

As Goodwin and others (Heath 1986; Kendon 2004; Mondada forthcoming; Schegloff 1998) argue, the gesture comes with the speaking, it is not a separate track of communication. Sign language is not running *alongside* spoken language. Alongside their relation to speaking, as we have noted above, the gestures have emergent sequential properties to them akin to both the ordering of words and the turns taken in conversation. The pointing here is only half-made and becomes a touch of the nose instead of a firm or definite point. The run-up nevertheless makes what is coming appreciable, in that we see the pointing coming before we could discern what it could point at. Indeed, run-ups or pre-pointings can, and do, happen this way to allow us to tune-in on the visual field. Often, we see what is being pointed at before the pointing comes to its pause because we are ongoing involved in monitoring scenes and so may well be able to correctly anticipate what the pointing is supposed to land on. Not by careful inspection of the direction of the finger, like the sight of a sniper's rifle, but by picking out the misbehaving vehicle ahead. The error of idea that we follow the finger is apparent when we try to point out a star in the night sky to a friend. The pointing finger almost always fails us when what is being pointed at cannot be picked out through a phrase or vague gestures in its direction.

Pointing as an embodied visual practice has been written about extensively elsewhere (Goodwin 2003; Hindmarsh and Heath 2000) and for that reason, along with its pertinence to pre-cognition, what I would like to do is move on to what immediately follows the pointing:

((*His finger then touches nose*))
P: I thought he was gonna::, ((*looks into car as they pass it*)) she was gonnae pull out and go for it

After the pivot of the [pointing *at*] into [*touching* nose], the latter gesture, in relation to the former, becomes a re-appraisal of what was seen as underway already. Noticeably cut-off in terms of how long we expect a point to last, the pointing's dissolution negates the thing that was now only seemingly seen. If it were a finger touching an object to make it relevant, it's a touch that is visibly withdrawn. The words that follow, 'I thought', could be taken as a report on what was occurring inside the brain of the passenger and that's the literal way 'thinking' is all too often dealt with by psychologists and neuroscientists. When having their brain scanned by PET or fMRI subjects are asked to 'think' of a number for instance. From a number of ordinary language philosophers the rejoinder to such experimental suppositions is that the uses of 'I think' and 'I thought' are *polymorphous* (Coulter 1989; Wittgenstein 1953).

In this driving event, to say 'I thought' is to make available what the passenger was *supposing* or *assuming* in making their gesture. It is the kind of use of 'I thought' that we say *after the fact* when indeed things may have gone wrong or have been revealed to not have been what the person has assumed or supposed. From the way the car ahead was moving, the passenger assumed they would pull out in front of them. An extreme example would be a police officer having accidentally shot a suspect saying 'I thought he was reaching for a gun' (from Bennett and Hacker 2003). Brains do not make such suppositions, police officers do. It is the police officer as a police officer that is held accountable not their brain. In neither case is there the need to suppose that an additional privileged process (e.g. thought) was going on. Moreover the phrase 'I thought' projects what will follow as an account for the mistaken gesture that immediately preceded. There's nothing essential about the gesture in directing the driver's attention, the passenger could have called out 'watch out!' or made one of those, often heard from passengers, sharp intakes of breath.

Concluding Remarks

The introduction of neuro-psychological theories of non-cognition and pre-cognition into geography by Nigel Thrift was one solution to a problem which continues to be rife in not only psychological studies of driving fifty years after Ryle's devastating critique, but in mobility studies more generally, which is the over-emphasis on intellectual forms of life and the attribution of its characteristics to travellers, drivers, passengers, tourists and more. The problem, overly simply put, is that what mobile actors are taken to be doing when they are reasoning in and about their movement is isomorphic with what theorists are doing when they theorise about mobility. Or as Thrift pithily puts it: 'Probably 95 percent of embodied thought is non-cognitive, yet probably 95 percent of academic thought has concentrated on the cognitive dimension of the conscious "I"'. (2000, 36). However the solution is not to import pre-cognition and cognition, thereby inadvertently smuggling a Cartesian division back into the analysis of mobile practices involving humans and new and old technologies.

What I have tried to do here instead is follow the example of a number of other philosophers and post-cognitive psychologists (Edwards 1997; Potter 2000) by returning to a mobile setting where we find 'thought' as an ordinary word being spoken by a passenger to a driver, where something is being seen to be so for a moment, then realised not to be so and initially gesturally repaired. I have pursued how this seemingly precognitive moment in driving or 'passengering' can be analysed without recourse to involuntary mental processes. Hopefully it gives a flavour of how we might begin to study actual instances of quite ordinary driving without falling into the cognitive theory's billiard ball game of mental causation. As private cognitive processes 'wanting', 'intending' or 'deciding' are taken to be indirectly accessible *causes* of actions or movements. To refuse to accept volition, intention or decision-making as specific mental processes is not to say that we do not slam on the brakes as a driver or find ourselves pushing the floor with out foot as a passenger *because* we intend to, want to or decide to do so, it is simply that the 'because' is not *causal*. If it were causal, once the passenger from the transcript had formed an intention to point out the car that was on a collision course he could simply relax and let the 'intention' *cause* his finger to rise and point toward the car. As a consequence, it is unwise to chart the redistribution of intentionality between driver and software or passenger and their unconscious because there is no 'thing' to be redistributed.

That what the passenger did was intended and that it was not involuntary (e.g. pre-cognitive) is covered by the fact that, even though the passenger was acting quickly, he was aware of what he was (intentionally) doing. If he was unaware of the fact that he had raised his arm and made a gesture that looked a lot like pointing at that car, *then* it would have been un-intended. It has nothing to do with a mental entity causing, or *not* causing, the action. This point applies to many of the other actions involved in driving that are gathered together as 'automatic', the prime example being changing gear (Groeger, 2000). That a driver grasps the gear lever in a certain way 'automatically' as they change gear may not be intentional and Thrift's remarks on ergonomics have some purchase there, but the driver's gear-changing is intentional. When some form of cruise control does the gear changing the attribution of intention changes. The driver, for instance, intentionally activating the cruise control and then a Latourian (1996) moment of displacement because the company that designed, engineered and programmed the cruise control intended the cruise control to behave in certain ways. What defines intended action is that the actor can provide reasons for what they are doing, or have done, which if you recall, is what the passenger provides, having suppressed his point toward the car. To say 'I thought' is to provide justification or excuse for our actions, not to name an internal state. An 'involuntary' gesture would be a nervous tic and one that others would not assume was intentional or that they meant anything by.

Here what we have seen is someone taking responsibility for their mistaken actions, rather than doing them either freely or under constraint. What the passenger's 'I thought' does is to take responsibility. 'He must mean what he gestures' to adapt a phrase from Cavell (2002). In examining what is excusable as the involuntary, Austin (1956-7) brings out that simple oppositions do not hold, quite the opposite, an action can be impulsive and intentional, or, an action can be intentional though not deliberate. Austin uses the amusing and potentially fatal example: while walking along the cliffs, on impulse pushing you over the edge. I both intended to do it while acting on impulse, yet what I did not do was pause to reflect on whether I should or not. In the passenger's case they intentionally point out the other car, there is no time to pause in advance and reflect on whether they should or not.

Reactions to impending dangers during driving are fascinating because they are at the edge of the involuntary. The mistake is to extend them across all driving and thus license grand statements about human psychology or the pre-cognitive nature of automobile systems when these rapid responses are relatively rare. Ultimately then I would argue there is no need for the pre-cognitive in understanding or analysing driving nor mobility in its more varied manifestations.

Acknowledgements

Barry Brown and Hayden Lorimer without whom this chapter could not have been written. Also analytic input from Ignaz Strebel and Lorenza Mondada. Commentaries from Mike Crang, Paul Harrison, Ben Anderson, Joe Painter, Paul McIlvenny and Ole Jensen. The drivers and passengers who have participated in the Habitable Cars project. ESRC Res 000-34-9758.

References

Austin, J.L. (1956–1957), 'A Plea for Excuses'. *Proceedings of the Aristotelian Society* (transcribed into hypertext by Andrew Chrucky), http://www.ditext.com/austin/plea.html.

Bennett, M.R., and Hacker, P.M.S. (2003), *Philosophical Foundations of Neuroscience* (Oxford: Blackwell).

Cavell, S. (2002), *Must We Mean What We Say?* (Updated Edition) (Cambridge: Cambridge University Press).

Coulter, J. (1983), *Rethinking Cognitive Theory* (London: Macmillan).

Coulter, J. (1989), *Mind in Action* (Atlantic Highlands, NJ: Humanities Press International).

Cresswell, T. (2006), *On the Move* (London: Routledge).

Edwards, D. (1997), *Discourse and Cognition* (London: Sage).

Goodwin, C. (2003), 'Pointing as Situated Practice', in S. Kita (ed.), *Pointing: Where Language, Culture and Cognition Meet* (Mahwah NJ: Lawrence Erlbaum), 217–41.

Groeger, J.A. (2000), *Understanding Driving: Applying Cognitive Psychology to a Complex Everyday Task* (Hove, UK: Psychology Press).

Hacker, P.M.S. (1996), *Wittgenstein: Mind and Will: An Analytical Commentary on the Philosophical Investigations, vol. 4.* (Oxford: Blackwell).

Hannam, K., Sheller, M. and Urry, J. (2006), 'Editorial: Mobilities, Immobilities and Moorings'. *Mobilities*, 1, 1–22.

Heath, C. (1986), *Body Movement and Speech in Medical Interaction* (Cambridge: Cambridge University Press).

Hindmarsh, J., and Heath, C. (2000), 'Sharing the tools of the trade; the interactional constitution of workplace objects'. *Journal of Contemporary Ethnography*, 29(5): 523–62.

Katz, J. (1999), *How Emotions Work* (London: University of Chicago Press).

Kendon, A. (2004), *Gesture: Visible Action as Utterance* (Cambridge: Cambridge University Press).

Larsen, J., Urry, J., and Axhausen. (2006), *Social Networks and Future Mobilities: Report to the UK Department for Transport* (Lancaster and Zurich: Lancaster University and IVT, ETH).

Latour, B. (1996), *Aramis, or the Love of Technology* (translated by C. Porter) (London: Harvard University Press).

Laurier, E., Lorimer, H., Brown, B., Jones, O., Juhlin, O., Noble, A., Perry, M., Pica, D., Sormani, P., Strebel, I., Swan, L., Taylor, A.S., Watts, L. and Weilenmann, A. (2008) 'Driving and passengering: notes on the ordinary organisation of car travel'. *Mobilities*, 3, 1–23.

Lerner, G.H. and Raymond, G. (forthcoming), 'On the practical re-intentionalization of action in interaction: midcourse pivots in the progressive realization of manual action', in G. Raymond, G.H. Lerner and J. Heritage (eds), *Enabling Human Conduct: Naturalistic Studies of Talk-in-Interaction in Honor of Emanuel A. Schegloff* (Amsterdam: John Benjamins).

Livingston, E. (2006), 'Ethnomethodological studies of mediated interaction and mundane expertise'. *The Sociological Review*, 54(3): 405–77.

Lupton, D. (2002), 'Road rage: drivers' understandings and experiences', *Journal of Sociology* 38, 275–90.

Michael, M. (1998), 'Co(a)agency and the Car: Attributing Agency in the Case of the "Road Rage"', in B. Brenna, J. Law and I. Moser (eds), *Machines, Agency and Desire* (Oslo: TVM), 125–41.

Miller, D. (2001), 'Driven Societies', in D. Miller (ed.), *Car Cultures* (Oxford: Berg), 1–33.

Mondada, L. (forthcoming), 'Deixis spatiale, gestes de pointage et formes de coordination de l'action', in J.-M. Barberis and M.C. Manes-Gallo (eds), *Verbalisation de l'espace et cognition situee: la description d'itineraires pietons* (Paris: Editions CNRS).

Ochs, E. (1979), 'Transcription as Theory', in E. Ochs (ed.), *Developmental Pragmatics* (New York: Academic Press), 43–71.

Pooley, C., Turnbull, J., and Adams, M. (2006), *A Mobile Century: Changes in Everyday Mobility in Britain in the Twentieth Century* (Aldershot: Ashgate).

Potter, J. (2000), 'Post-cognitive psychology'. *Theory & Psychology*, 10(1): 31–7.

Potter, J. (2005), *Paper presented at ESRC Neuroscience, Identity and Society Seminar Series.*

Potter, J. and te Molder, H. (2005), 'Talking cognition: mapping and making the terrain', in J. Potter and H. te Molder (eds), *Conversation and Cognition* (Cambridge Cambridge University Press), 1–54.

Schegloff, E.A. (1998), 'Body torque'. *Social Research*, 65, 535–86.

Schegloff, E.A. (2007), 'A tutorial on membership categorization'. *Journal of Pragmatics*, 39, 462–82.

Sheller, M. and Urry, J. (2006), 'The new mobilities paradigm'. *Environment and Planning A*, 38, 207–26.

Sormani, P. (2004), 'Order in disorder: some ethnographic observations in traffic'. *Copies available from the author, Department of Sociology, University of Manchester.*

Thrift, N. (1996), *Spatial Formations* (London: Sage).

Thrift, N. (2000), 'Still Life in Nearly Present Time: The Object of Nature', *Body and Society,* 6(3–4): 34–57.

Thrift, N. (2004a), 'Driving in the City'. *Theory, Culture and Society*, 21(4/5): 41–59.

Thrift, N. (2004b), *Knowing Capitalism* (London: Sage).

Thrift, N. (2005a), 'But malice aforethought: cities and the natural history of hatred'. *Transaction of the Institute of British Geographers*, 30, 133–50.

Thrift, N. (2005b), 'From born to made: technology, biology and space'. *Transactions of the Institute of British Geographers,* NS 30, 463–76.

Thrift, N. and French, S. (2002), 'The Automatic Production of Space'. *Transactions of the Institute of British Geographers,* 27, 309–35.

Urry, J. (2000), *Sociology beyond Societies: Mobilities for the Twenty-First Century* (London: Routledge).

Urry, J. (2003), 'Social networks, travel and talk'. *British Journal of Sociology*, 54(2): 155–75.

Urry, J. (2004), 'Connections'. *Environment and Planning D: Society and Space*, 22, 27–37.

Watson, R. (1999), 'Driving in Forests and Mountains: A Pure and Applied Ethnography'. *Ethnographic Studies*, 3, 50–60.

Watson, R. (2003), 'The Anthropology of Communication: Foundations, Futures and the Analysis of "Constructions" of Space', in T. Lask (ed.), *Constructions sociales de l'espace, Les territoires de l'anthropolgie de la communication* (Liège: Les Éditions de l'Université de Liège), 193–204.

Wittgenstein, L. (1953), *Philosophical Investigations* (translated by G.E.N. Anscombe) (Oxford: Blackwell).

Chapter 6

Flying: Feminisms and Mobilities – Crusading for Aviation in the 1920s

Dydia DeLyser

After earning her pilot's license in 1928, Louise Thaden detailed what she called 'the soothing splendor of flight' amid the high-pressure atmosphere of airborne competition to reach the limits of speed, distance, altitude, and endurance. Describing the power of aerial perspective as well as the sensations and personal meanings of flying, she wrote: I had 'the ability to go up into God's heaven, to look toward distant horizons, to gaze down upon the struggling creatures far below, to forget troubles which so short a time before seemed staggering, just to feel the lifting of the wheels from the ground, to hear the rush of air past the cabin window, to squint into the sun, toying with the controls, to feel the exhilaration of power under taut leash, responsive to whim or fancy, to feel, if only for one brief moment, that I could be master of my fate...' (Thaden 2004 [1938], 65). For Thaden, as for other women in the late 1920s and early 1930s raised under rigid gender norms and restricted mobility, flying was a practice of liberation.

It is the voices and practices of women like Thaden that this chapter seeks to breathe into scholarship on women in aviation, on gendered mobilities, and on geographies of practice. I focus here on three white American women, Louise Thaden, Ruth Nichols, and Amelia Earhart – professional pilots who spoke publicly, wrote about aviation and penned autobiographies (Earhart 1977 [1932]; Nichols 1957; Thaden 2004 [1938]). Among the most accomplished women pilots of their time, they earned Air Transport Pilot's ratings (the highest aircraft license), held records in speed, endurance, altitude, and aerial refueling – records they often swapped with each other – and they competed for first flights, like Earhart's solo Atlantic crossing, to which she beat Nichols by only a matter of days.[1] Though Nichols was from a wealthy background and had no difficulty paying for her training, Thaden and Earhart had lower-middle-class backgrounds and took jobs to earn their wings – Earhart at the phone company, Thaden selling coal (Earhart 1977 [1932]; Nichols 1957; Thaden 2004 [1938]; see also Brooks-Pazmany 1938).[2] All three became relatively successful professional pilots, supporting at

---

1 Nichols had crashed on a previous Atlantic attempt in 1931, and at the time of Earhart's flight was in final preparations for a second attempt (Nichols 1957, 208–10).

2 Thaden sold coal, fuel oil and building materials for two years; her boss was a stock holder in Beech, and when he learned of her strong aviation interest arranged for her

least themselves (though not often their families) with their flying. Significantly, all three were articulate about what it meant to be a woman pilot in the late 1920s and early 1930s.

Before delving into their experiences, I briefly outline how feminist understandings of embodiment and geographies of practice offer new possibilities for thinking through what has been identified as a 'mobilities turn' across the social sciences and humanities. I then attempt to draw that scholarship forward to talk about what those mobilities could actually be used for – about what the point of being a woman pilot in 1929 really was. My purpose is first, through qualitative historical research on flying, to draw women's experiences into scholarship on mobilities. And second, to marshal that understanding in order to enrich contemporary chronologies of feminist political activism in the United States that have remained fixed upon the suffrage movement and its victory in 1920, leaving a gap between then and the second-wave feminism of the 1960s and 1970s – a gap often presumed devoid of feminist activity and action (Ware 1993).

Moving Bodies

As Lise Nelson and Joni Seager (2005, 7, 3) point out 'Women's lives are so easily and so often trivialized and 'disappeared' that a commitment to taking women seriously needs conscious and continuous reassertion,' so that the 'work of making – and keeping – women's lives visible is [still] far from complete, and such projects remain at the heart of feminist geography.' But the lives of women in the past must not be left out of that project. Nevertheless, because, as Mona Domosh and Karen Morin (2003, 260, 262) note, much feminist geography 'remains resolutely non-historical, and most historical geography fails to consider women, or gender, or difference … writing women's lives, voices, stories, and experiences into historical geography … remains a pressing issue.' While feminist geographers have anchored their work in embodied experience(s) (see Nelson and Seager 2005), cultural geographers too have sought to understand socio-spatial life in ways that, as Hayden Lorimer has put it (2005, 83), 'better cope with our self-evidently more-than-human, more-than-textual, multisensual worlds.' We have reached for new ways of understanding that allow for the 'transient aspects of living,' the embodied, ephemeral, and mobile practices and performances of everyday life (Lorimer 2005, 83). Meanwhile, Jane Jacobs and Catherine Nash (2003, 276) point out that 'considering embodiment as *not only, but also always* the product of representation, regulation, relationality, and performative reiteration … is overtly concerned with the implications of attending to non-discursive and bodily materiality for those [like women] historically subordinated by discourses about their (supposed) innate inferiority.' Together these geographers point toward

---

to work in Beech's west coast office and for her to learn to fly as part of that job, beginning in 1927 (Thaden 2004 [1938]).

new ways to understand how human movements are socio-spatially understood – for issues of mobility, of course, are entangled with meaning and power, with issues of 'race,' class, and gender, and, not insignificantly, with both representation *and* practice. It's what Tim Cresswell (2006, 264–5) has termed the 'social baggage that accompanies' differential mobilities, for mobility must be viewed as more than 'getting from A to B,' and seen to include the ways that different mobilities and different kinds of mobilities can be practiced, contested, represented, and understood, leading to richer understandings of the social dimensions of the ways we move, and the socio-spatial meanings these mobilities are, and have been, given.

Across the social sciences and humanities, invigorated interest in mobility has sparked what Mimi Sheller and John Urry term a 'new mobilities paradigm,' bringing together social and transportation research to understand how the 'spatialities of social life presuppose (and frequently involve conflict over) both the actual and the imagined movement[s] of people' (Sheller and Urry 2006a, 208). Advocates of this 'mobility turn,' as Cresswell (2005, 448) notes, argue that 'contemporary society and culture should be understood through its modes of mobility as much as by its spaces and places.' Mobilities scholars seek to move from a 'sedentarist' world view that prioritizes and valorizes fixity and stability, to a 'nomadic' one focused on mobility and fluidity (Cresswell 2006; Sheller and Urry 2006a). Much of this research has been contemporary, focused, as Sheller (2004) has pointed out, on very recent developments in transportation and technologies, on spaces of flows emergent in an increasingly more rapidly mobile world. Of particular interest have been spaces perceived as anonymous, the 'non-places,' like the large international airports perceived as 'iconic space[s] for discussions of modernity and postmodernity' (Cresswell 2006, 220; see also Gottdeiner 2001; Pascoe 2001; Merriman 2005; Adey et al. 2007; Adey 2008c).

The challenge has been, as Cresswell (2006, 255) has pointed out, that the 'general celebration of the nomadic in contemporary theory too often levels out agency so that … differences in the experience of mobility disappear.' Broad, sweeping accounts have given little pause to detailed empirical concerns and the nuances of practice. As Linda McDowell (2005, quoting Adkins 2004, 146) has cautioned, 'the privileging of mobility and mobile subjects [risks] 'reinstall[ing] and idealis[ing] a disembodied, disembedded subject who moves unfettered across and within the social realm;' a subject that typically is masculine…,' in work that pays little heed to the struggles inherent in many people's everyday practices of mobility. In fact, feminist scholars have urged caution from a romantic reading of mobility (Kaplan 2006) grounded in a 'bourgeois masculine subjectivity' (Skeggs 2004, 48) that overlooks the ways that issues of 'race,' class, gender, and difference have profound impacts on mobilities, leaving some – often women – mobility impaired (Sheller and Urry 2006a), for mobility is 'a gendered activity that is often more available to men than it is to women,' and that is often very differently understood and experienced by women than by men (Cresswell 2005, 448; Uteng and Cresswell 2008). Mobility is an 'uneven resource,' and mobilities scholars seek not to privilege a mobile subjectivity, but to track 'the power of discourses and practices of mobility in creating both movement and stasis' (Sheller and Urry 2006a, 211).

Why Flying?

To J. Nicholas Entrikin and John Tepple (2006), many geographic studies of practice have lost understandings of individual agency – of meaningful action. I hope to address that, in the contexts of the above debates, by examining how flying *works* as a gendered and embodied practice of mobility – as action with agency – and by situating women's flying in the historical context of postsuffrage feminism. I aim to show what these practices mean, and what these mobilities were used for.

Looking at flying is not new in geography (Light 1935; Vowles 2006; Adey 2008a). But, despite a surge of works in transportation geography since the 1970s on airports, air transportation, air travel, airlines, aircraft, and air traffic, there has been no attention to flying itself, or to pilots as flyers – geographers seem to have overlooked the embodied experiences and meanings of flying (Vowles 2006; Adey 2008a, 2008c). Indeed, as Saulo Cwerner (2006, 194) notes, 'the growing literature on aviation in the social sciences has to a large extent focused on the social and cultural analysis of airports, highlighting only selective aspects of air mobility.' While such contributions remain valuable, they overlook the embodied, empirical significance of flying as practice, and neglect the experience of piloting that forms so significant a part of flying for those who genuinely engage it.

Something similar can be said for recent works on mobilities: while scholars have addressed, for example, issues of air power, movement through airports, long-distance commuting by air, and urban commuting by helicopter, and others have written on the socio-spatial meanings of walking and of driving, little has been written on the embodied practice of flying (for example, Gottdeiner 2001; Pascoe 2001; Featherstone et al. 2005; Merriman 2005, 2007, 2009; Wylie 2005; Kaplan 2006; Lassen 2006; Cwerner 2006; Adey 2006, 2007, 2008b). Beyond a lack of attention to flying-as-doing, what work there is on travel by air often describes it as merely a transportation necessity (Lassen 2006). What such work overlooks in its inattention to flying, is not just one other means of locomotion, or one other means of understanding embodied experience, but what purpose such active practice actually serves. What such mobilities can actually be used for.

But if flying, for those who don't actually do it, now seems pedestrian, that has not always been the case. At a time, in the last century, when women in the US had only recently won suffrage, when most were still expected not to work outside the home, and when their very movements away from home were often sharply curtailed – by, for example, their clothing, their lack of access to paid employment, and by societal expectations – women who learned to fly described the ecstasy and empowerment of that experience and, in their very embodied practices of flying, helped reconfigure the gendered moral geographies of their time and transform the 'geography of expectations' that shrouded men's and women's mobilities (Cresswell 2005, 448). For women in the late 1920s and early 1930s, the experience of flying 'belied the rhetoric of domesticity,' and women aviators wrote articulately about their experiences (Corn 2002). Aviation journalist and flyer Margery Brown spelled it out: through flight, women could achieve liberation, both physical and metaphorical, 'Women are seeking freedom. Freedom in the skies! They are soaring above ... [the obstructions to] their sex which have kept them earth-bound. Flying is a symbol of freedom from limitation' (quoted in Oakes 1985, 4).

On Women in the History of Aviation

In fact, since its launch in 1903, powered flight has been a male-dominated endeavor – long understood as a masculine undertaking, and once seen as the exclusive province of men. Orville Wright, for example, rejected all women applicants from his flying school (Brooks-Pazmany 1983). At a time when the women's sphere was defined as the private space of the home, flying, for women was seen as socially inappropriate, women as pilots were thought 'temperamentally unfitted,' creatures for whom flying, it was thought, might even be physically impossible (Claude Graheme-White quoted in Moolman 1981, 9; Brooks-Pazmany 1983).

As Ruth Law, a record-breaking pilot licensed in 1912 wrote, 'There is the world-old controversy that crops up again whenever women attempt to enter a new field – Is woman fitted for this or that work?' (Law 1918, quoted in Moolman 1981, 33). At a time when newspapers ran headlines – as the Detroit *Free Press* did as late as 1911 – 'Ought women to aviate?' (quoted in Moolman 1981, 9) women pilots, by actively pursuing flight, and by doing so in the face of often-strong opposition, 'enlarged the traditional bounds of a woman's world' and radically redefined women's mobilities (Moolman 1981, 7).

Though the first women earned their licenses in the early 1910s, during WWI women were prohibited from flying in the military and civilian flying was curtailed, effectively curbing women's flying. Then, stories of heroic male 'Aces' gave pilots a reputation for daring, while the War's end brought home a generation of unemployed male pilots and a wealth of surplus aircraft. Many male pilots began earning their living as 'barnstormers' traveling the country performing their aerial prowess in flying circuses of near-ground stunts, and selling rides to earn a living (Brooks-Pazmany 1983; Corn 2002). A few women joined the ranks of the barnstormers, itinerant performers who brought death-defying airborne feats to airports and open fields across the US. But while barnstorming provided income for pilots, periodic accidents and constant perceived risks associated with both the aircraft and the reputations of the pilots themselves worked against the acceptance of aviation by the public (Brooks-Pazmany 1983).

Though flying itself was no longer new, in the 1920s it possessed, if anything, even more prophetic promise and romantic excitement for spectators and pilots alike – but flying, for anybody, was a rarity (Corn 2002; Wohl 2005). US commercial aviation had been ignited by federal airmail contracts in 1925 but by 1929 most airlines still relied on those mail contracts for revenue, and scheduled passenger service was virtually non-existent. As Ruth Nichols (1957, 63) put it, talking about her 43-state tour in a Curtiss Robin in 1929, though few had flown, 'People everywhere ... were fascinated with the idea of flying.' By the late 1920s, aircraft were becoming a reliable means of long-distance transportation, but acceptance of air travel remained limited. Indeed, as Corn notes, while the public was 'enthusiastic about airplanes and about flying in the abstract ... in great numbers they refused to fly,' and industry insiders commonly complained that it was 'fear not fare' that kept the public on the ground (2002, 74).

Aviation had made great strides on the drawing boards, but to the general public it was remote; aviation-in-the-doing, flying as a practice, was nearly non-existent. To make aviation more acceptable it had to become more ordinary, perceived as safe, and easy. And it had to be done, practiced, engaged, by 'regular' folk not just military Aces. Aviation historians agree that the foundational obstacle was the public's perception of aviation as risky, and pilots as a so-called 'breed apart' (Brooks-Pazmany 1983; Corn 2002). Despite the vast promises flying held for speed and ease of communication and transportation in a modernist-era optimistic future, the dominant image of the male flying 'Ace' was hurting aviation (Corn 2002; Wohl 2005). With only industry assertions that flying was safe, easy, and everyday, the public did little in response. Hearing or reading about flying was very different from doing it and seeing it done. So it was the active practice of flying by specific individuals that persuaded the public that aviation was not just intriguing, but also safe and doable (Corn 2002).

By the late 1920s the woman pilot became aviation's antidote to the risk-taking ace, proving-by-doing that aviation was both safe and straightforward. Where gendered stereotypes of women as fragile, timid, and unmechanical, along with lingering Victorian understandings of gendered spaces (where the home was the province of women) had long hampered women explorers, travelers and women automobilists (see Scharff 1992; Morin 1998, 1999), women pilots across the country now made new opportunities *because* they were women. They found work promoting and selling aircraft, advertising airlines, and drawing attention to aviation-related products. Where gender stereotypes had previously kept many women out of the cockpit, now women turned those stereotypes in on themselves, earning their first aviation jobs because of the very stereotypes that had for so long impaired their mobilites, and downcast their abilities. As Louise Thaden wryly put it, 'Nothing impresses the safety of aviation on the public quite so much as to see a woman flying an airplane. If a woman can handle it, the public thinks it must be duck soup for men' (quoted in Corn 2002, 75). Ruth Nichols, herself a veteran of four nearly fatal accidents, agreed: the public thought 'it must be easy,' she said, if the so-called 'fragile sex' can do it (quoted in Corn 2002, 76).

Women Take to the Air

After Charles Lindbergh's acclaimed New-York-to-Paris flight in 1927, Amelia Earhart's trans-Atlantic flight as a passenger in 1928 and the first Women's Transcontinental Air Race of 1929 sparked women's interest in aviation. In January of 1929 there were only twelve licensed women pilots in the US (Earhart 1977 [1932], 146). By year's end that number had increased tenfold (Nichols 1957, 97). By 1932 there were more than 500 (Earhart 1977 [1932], 146), amounting to what one commentator called a feminine 'rush to the cockpits' (quoted in Corn 2002, 73).

A rush it may have seemed, but it was not unobstructed. As Thaden described it, 'Being, or trying to be, a woman pilot' in those days 'had tremendous built-in disadvantages. Basically we were usurpers in a man's exclusive world. The penalties were severe in the never-ending struggle to accumulate the flying time which alone could develop piloting skills ... [W]e found, as anticipated, that job openings were almost nonexistent. As was comparably true with other industries, a high wall of prejudice securely enclosed us' (Thaden 2004 [1938], xii–xiii). Despite the obstacles, women pilots persevered. And, once they gained their licenses, they labored to make it easier for other women to do the same. As Thaden put it, 'Most of us who had gotten a foot wedged in the door conducted a relentless personal crusade to increase its opening, enabling other women to enter this sacred portal' (Thaden 2004 [1938], xiii).

Some of these women flew alone – it was not uncommon in those days for women, like Nellie Willhite in South Dakota, to be the only woman in their towns, cities, states, or even regions to earn pilot's licenses. Still, even Willhite participated in the collective efforts of women pilots, becoming, in 1929, a Charter Member of the Organization of Women Pilots, and founding the South Dakota Chapter though she was at that time its only member (Brooks-Pazmany 1983). Others found the camaraderie of women pilots at their local airports, while still others, Nichols, Thaden, and Earhart among them, demonstrated women's mobility on a national scale, competing in races, attempting numerous records, and establishing 'first flights.'

At a time before the public had become satiated by first flights and record attempts,[3] aviation for women was a hive of activity where a relatively small group of professional pilots, sponsored by various companies, labored to bring aviation, and women in aviation in particular, into the public eye by setting new records or being the first woman to accomplish a particular feat (Brooks-Pazmany 1983; Corn 2004). For all the discrimination they faced, they were also able to use negative gender stereotyping to their advantage since, as Thaden explained, 'Women pilots were oddities and therefore generally more 'newsworthy' than were the male counterparts' (Thaden 2004 [1938], xii–xiii). As Amelia Earhart put it, '[in 1929] the returned aviator still rated spectacular headlines. He was front page

---

3 Nichols, *Wings for Life* puts this date around 1932 on p. 219.

"news," *she* was even "front-pager"' (Earhart 1977 [1932], 87). Nichols, known as the 'flying debutante' for her high-society upbringing, explained the conundrum she faced after pioneering a New-York-to-Miami route: 'Until now I had feared publicity like smallpox. It was a family fetish that nice people didn't get their names in the papers, except possibly in the society columns on the occasion of an engagement or a wedding. But ... [now] I gained a new perspective. Publicity had a definite and important business value to the infant aviation industry. As one of the nation's few women flyers, I had become a valuable publicity asset' (Nichols 1957, 49).

By the late 1920s, with a few jobs available for women in aviation, some were now able to support their flying habits. Amelia Earhart became Aviation Editor at *Cosmopolitan Magazine*, as well as Assistant General Traffic Manager for Transcontinental Air Transport (later TWA); Louise Thaden became a sales representative and demonstration pilot for Travel Air; and Ruth Nichols promoted 'Aviation Country Clubs.' They endorsed products as well: Earhart appeared for Lucky Strike cigarettes as well as for her own line of clothing and luggage, Thaden and Nichols both endorsed Kendall Oil.[4]

But while their record setting was competitive by nature, it was also more broadly collaborative, expressly intended to advance both women and aviation simultaneously. As Nichols – the only person to hold records for speed, altitude, and distance concurrently – observed, 'Records are made to be broken, and I only wish that more girls could get good ships and keep setting new marks all the time. It has long been my theory that if women could set up some records, in many cases duplicating the men's, the general public would have more confidence in aviation' (quoted in Brooks-Pazmany 1983, 27). Indeed, even though women were involved in establishing airfields and operating aviation businesses, because those were multi-person ventures (usually involving men as well) and not often seen as glamorous, for women to put themselves and aviation in the spotlight they had to be setting records or first flights (Brooks-Pazmany 1983).

But, because women's flying had once been considered inappropriate, and because few pilots were women, women's records were initially not officially recorded by US or international monitoring agencies. Unsatisfied with this lack of recognition, women pilots, eager to attract attention to women's accomplishments as well as to aviation, lobbied, and in 1929 gained recognition for their records in a separate women's record category. Their victory in establishing the women's category served as a tremendous impetus for women pilots to make record attempts

---

4   For general information about women pilots in this period see Brooks-Pazmany 1983, especially p. 15–16. Earhart wrote for *Cosmopolitan* in 1928; her clothing and luggage line appeared later and details are in the George Palmer Putnam Collection of Amelia Earhart Papers at Purdue University; Nichols describes her work with Aviation Country Clubs in Nichols 1957; ads featuring Thaden and Nichols endorsing Kendall oil (and other products) in the author's collection.

– for fame, for their backers, for aviation, and for women (see Brooks-Pazmany 1983).

Crusading for Feminism

Understanding women's airborne mobilities must not stop here. We must also endeavor to understand what these gendered mobilities were actually used for. Women pilots in the late 1920s saw themselves as 'pioneers,' and worked deliberately to set examples for other women to follow. Ruth Nichols (1957, 52) put it this way:

> Back in 1929 our hopes were high, our enthusiasm unbounded and our energies almost limitless. [A]ll of us were dedicated disciples to the widening new frontiers of the air, and we threw ourselves into the work with the energy of devotees to an ideal. Through flight came joy, new perspectives, and help in blazing new frontiers of the mind and heart for the people of the earth.

As Thaden (2004 [1938], 51) put it, 'We women pilots were blazing a new trail.' And, though they flew for themselves, they also labored towards shared goals. 'Above all else,' wrote Thaden (2004 [1938], 100) 'we felt the responsibility of doing a good job, for upon our shoulders rested the fate of future employment of women pilots.' And as Earhart told Thaden in 1929, 'We women pilots have a rough, rocky road ahead of us' … 'Each accomplishment, no matter how small, is important. Although it may be no direct contribution to the science of aeronautics nor to its technical development, it will encourage other women to fly. The more women who fly, the more who become pilots, the quicker will we be recognized as an important factor in aviation.' To which Thaden answered, 'You can count on me to do everything possible to help' (quoted in Thaden 2004 [1938], 147).

This point about flying was never far from their minds. But it was perhaps Earhart, the only one with a full-time publicist, who made the strongest impression. Giving literally hundreds of speeches each year about women in aviation, appearing live and on radio, writing a column about women in aviation, books about her flights, and personally answering thousands of letters from fans (see Ware 1993). As Earhart put it (Earhart 1977 [1932], 161), 'Records as such may or may not be important, but at least the more of them women make, the more forcefully it is demonstrated that they can and do fly. Directly or indirectly, more opportunities for those who wish to enter the aviation world should be opened by such evidence.'

These motivations led women pilots like Earhart, Thaden, and Nichols to become what scholars have since termed 'liberal' or 'social' feminists working for the cause of women in ways quite different from the feminists of the previous generation. Coming of age in an era after suffrage had been won, women pilots of the late 1920s and early 1930s sought to distinguish themselves from what they (and others of those days) saw as the dowdy feminists of the generation before them. As Dorothy Dunbar Bromley wrote in *Harper's* magazine in 1927 (quoted in Latham 2000, 26), the term 'feminism' itself had by then become 'opprobrious,' as the new generation of women then in their 20s and 30s associated the term with the 'old school of fighting feminists who wore flat heels and had very little feminine charm.'

Earhart was perhaps most articulate about the liberal or social feminist model she helped establish. She understood that not only could she herself actively break gender stereotypes, that not only could she herself do things that other women had previously not thought possible, but that in the doing of those things, in her own practice of mobility, she herself would become a powerful role model for thousands of others, until one day, she believed, the weight of those limiting stereotypes would be defeated. In conversation with Thaden, Earhart spelled out her strategy. 'Just what *do* you want?' asked Thaden.

> Recognition for women. Men do not believe us capable. We can fly – you know that. Ever since we started we've batted our heads against a stone wall. Manufacturers refuse us planes. The public have no confidence in our ability. If we had access to the equipment and training men have, we could certainly do as well. Thank heaven, we continue willingly fighting a losing battle. Every year we pour thousands of dollars into flight training with no hope of return. A man can work his way through flying school, or he can join the army. When he has a license he can obtain a flying job to build up his time. A man can borrow the latest equipment for specialized flights or for records. And what do we get? Obsolete airplanes. And why? Because we are women …

'Well,' answered Thaden, 'there isn't much to be done about it, except to keep trying.' 'That's right,' replied Earhart. 'But if enough of us keep trying we'll get some place' (Thaden 2004 [1938], 148–9).

In recent decades, however, seen from contemporary perspective, this type of liberal or social feminism, and the period of feminist activity after suffrage in 1920, has often been mischaracterized by scholars as a time when the lack of a broad-based collective women's movement meant the lack of feminist achievement, and the lack of a feminist agenda (Ware 2003). Exploring the embodied mobilities of women aviators, though, reveals a richer understanding of women's mobilities and feminist activity in this period, but not because Earhart and the other women pilots were themselves so exceptional. Indeed, women pilots understood themselves as powerful role models for other women, women they hoped to inspire not just to fly, but more broadly to break the rigid gender roles limiting women's mobilities, women's social roles, and women's professional opportunities. They often (though not always) gained notoriety as individuals, but their words and deeds, when closer examined, reveal their collective efforts on behalf of women. Further, such individual attention should not obscure collective action, or the ability of one woman in the spotlight to inspire numerous others. As Earhart biographer

Susan Ware has pointed out (Ware 1993, 13), 'women like [these] personified a model of women's postsuffrage achievement that was widely, and sympathetically, reported in the media and spread by popular culture [at the time]. These popular heroines kept feminism alive in a period usually thought to be barren of gender consciousness ... and provide[d] ... a bridge between suffrage activism and the [emergence of second-wave feminism] in the 1960s and 70s.' At a time when there were few new models available for young women, 'desperate to break out of old patterns but not quite sure what to replace them with' women pilots, through the embodied practice of flying, showed what women could do, and how different women's mobilities and women's lives could be (Ware 1993, 35).

Conclusion

Aviation developed in the US through the plans of engineers and mechanics, but most powerfully for the public in the *doing*, in the actual flying, practices in which women aviators were an integral part (see Brooks-Pazmany 1983). Women pilots of the late 1920s and early 1930s crusaded for women and aviation, engaging flying as an active and embodied practice that demonstrated what broader roles for women in society could be, and served to mobilize other women to actualize those mobilities.

Women's embodied mobilities, in practice and in the representations of that practice – in their speeches, articles, and books – were, in fact, good for something. They served as mobile symbols of women's emancipation in the postsuffrage era (see Ware 1993). As Earhart put it (quoted in Ware 1993, 80), 'Someday, I daresay, women can be flyers and yet not be regarded as curiosities' for 'women are people in the air.' As women pilots carried their message of women's opportunities and abilities, women's equality and women's liberation across the country through their embodied practices of flight, they served as new role models. And they experienced that liberation, that freedom, themselves through their own exalted mobilities. As Louise Thaden put it (2004 [1938], 17), describing one of her earliest flights, 'The engine sang an even song. The airplane perfectly rigged. It [gave] a heady feeling of supremacy – a day good for the ego, one of those days when it's hard to fly right side up.' The experience could be intoxicating. As Thaden described it (2004 [1938], 56), 'Tingling, I thundered across the white finish line on Columbus Airport. For the first time I allowed myself to go beserk in the air, showing off ... – a vertical roll followed by a loop, coming out in a spin, kicking out, circling inside the field in a low vertical bank ... Smug, self-satisfied, I brought the ship in for a landing, purposely high as an excuse for a spectacular sideslip, a violent fishtail, to gently kiss the runway in a slow three-point landing.' But their mobilities were directed not only to themselves, but to all women. As pilot and journalist Margery Brown put it in 1929 (quoted in Ware 1993, 63), 'The woman at the wash-tub, the sewing machine, the office-desk, and the typewriter can glance up from the window when she hears the rhythmic hum of a motor overhead, and say, "If it's a woman she is helping free me too!" A victory for one woman is a victory for all ... a woman who can find fulfillment in the skies will never again need to live her life in some man's spare moments.'

Women pilots of the late 1920s and early 1930s, through their embodied practices of flying, challenged the dominant and restrictive spatialities of their day. Understanding such 'breaches,' as Bondi and Davidson put it (2004, 23) is one of the goals of feminist geography. But understanding too what those mobilities were actually *used for* reveals women pilots as intentional agents, not just actively practicing their mobilities, not just breaching dominant spatialities, but actively and intentionally *engaging* their mobilities in a crusade for women, for aviation, and for new forms of feminism in the postsuffrage era.

References

Adey, P. (2006), 'Airports and air-mindedness: Spacing, timing and using the Liverpool Airport, 1929–1939,' *Social and Cultural Geographies* 7(3): 343–63.

Adey, P. (2008a), 'Aeromobilities: Geographies, subjects, vision,' *Geography Compass* 2 (September), 1318–66.

Adey, P. (2008b), 'Architectural geographies of the airport balcony: Mobility, sensation and the theatre of flight,' *Geografiska Annaler Series B – Human Geography* 90B(1): 29–47.

Adey, P. (2008c), 'Airports, mobility and the calculative architecture of affective control,' *Geoforum* 39, 438–51.

Adey, P., Budd, L. and Hubbard, P. (2007), 'Flying lessons: Exploring the social and cultural geographies of global air travel,' *Progress in Human Geography* 31(6): 773–91.

Adkins, L. (2004), 'Gender and the Post-Structural Social,' in B.L. Marshall and A. Witz (eds) *Engendering the Social: Feminist Encounters with Sociological Theory* (Buckingham and Philadelphia: Open University Press), 139–54.

Bondi, L, and Davidson, H. (2004), 'Situating Gender' in J. Seager, K. Agot, M. Blake, L. Bondi, and L. Nelson, *Companion to Feminist Geography* (London: Blackwell), 15–31.

Brooks-Pazmany, K. (1983), *United States Women in Aviaiton 1919–1929* (Washington: Smithsonian Institution Press).

Corn, J.J. (2002), *The Winged Gospel. America's Romance with Aviation* (Baltimore: Johns Hopkins University Press).

Cresswell, T. (2005), 'Mobilising the movement: The role of mobility in the suffrage politics of Florence Luscomb and Margaret Foley, 1911–1915,' *Gender, Place, and Culture* 12(4): 447–61.

Cresswell, T. (2006), *On the Move: Mobility in the Modern Western World* (London: Routledge).

Cwerner, S.B. (2006), 'Vertical flight and urban mobilities: The promise and reality of helicopter travel,' *Mobilities* 1(2): 191–215.

Domosh, M. and Morin, K. (2003), 'Travels with feminist historical geography,' *Gender Place and Culture* 10(3): 257–64.

Earhart, A. (1977 [1932]), *The Fun of It: Random Records of My Own Flying and of Women in Aviation* (Chicago: Academy Press Limited).

Entrikin, J.N., and Tepple, J.H. (2006), 'Humanism and Democratic Place Making,' in S. Aitken and G. Valentine (eds) *Approaches to Human Geography* (Thousand Oaks, CA: Sage Publications), 30–41.

Featherstone, M., Thrift, N., and Urry, J. (2005), *Automobilities* (Thousand Oaks, CA: Sage Publications).

Gottdeiner, M. (2001), *Life in the Air: Surviving the New Culture of Air Travel* (Boston: Rowman and Littlefield).

Jacobs, J.M., and Nash, C. (2003), 'Too little, too much: Cultural feminist geographies' *Gender Place, and Culture* 10(3): 265–97.

Kaplan, C. (2006), 'Mobility and war: The cosmic view of US "air power"', *Environment and Planning A* 38, 395–407.

Lassen, C. (2006), 'Aeromobility and work,' *Environment and Planning A*, 38, 301–12.

Latham, A.J. (2000), *Posing a Threat: Flappers, Chorus Girls and other Brazen Performers of the American 1920s* (Middletown, CT: Wesleyan University Press).

Law, R. (1918), 'Let women fly!' *Air Travel*.

Light, R.U. (1935), 'Cruising by airplane: Narrative of a journey around the world,' *Geographical Review* 25(4): 565–600.

Lorimer, H. (2005), 'Cultural geography: The busyness of being "more-than-representational"', *Progress in Human Geography* 29(1): 83–94.

McDowell, L. (2005), 'Travelling times: Some skeptical comments on the idea of a new mobilities paradigm,' delivered at Institute of British Geographers meeting, London.

Merriman, P. (2005), 'Driving places: Marc Augé, non-places and the geographies of England's M1 motorway,' in M. Featherstone, N. Thrift and J. Urry (eds) *Automobilities* (London: Sage), 145–67.

Merriman, P. (2007), *Driving Spaces. A Cultural-Historical Geography of England's M1 Motorway* (Oxford: Wiley-Blackwell).

Merriman, P. (2009), 'Automobility and the geographies of the car', *Geography Compass*, 3, 586–99.

Moolman, V. (1981), *Women Aloft* (New York: Time-Life Books).

Morin, K. (1998), 'British women travelers and constructions of racial difference across the nineteenth-century American West,' *Transactions of the Institute of British Geographers* 23, 311–30.

Morin, K. (1999), 'Peak practices: Englishwomen's 'heroic' adventures in the nineteenth-century American West,' *Annals of the Association of American Geographers*, 89, 489–514.

Nelson, L., and Seager, S. (2005), 'Introduction', in *Companion to Feminist Geography* (London: Blackwell), 1–12.

Nichols R. (1957), *Wings for Life* (Philadelphia: J.B. Lippincott Company).

Oakes, C.M. (1985), *United States Women in Aviation, 1930–1939* (Washington: Smithsonian Institution Press).

Pascoe, P. (2001), *Airspaces* (London: Reaktion Books).

Rich, D.L. (1993), *Queen Bess: Daredevil Aviator* (Washington: Smithsonian Institution Press).

Scharf, V. (1992), *Taking the Wheel: Women and the Coming of the Motor Age* (Albuquerque: University of New Mexico Press).

Sheller, M. (2004), 'Mobile publics: Beyond the network perspective,' *Society and Space* 22, 107–25.

Sheller, M. and Urry, J. (2006a), 'The new mobilities paradigm,' *Environment and Planning A* 38, 207–26.

Sheller, M., and Urry, J. (2006b) *Mobile Technologies of the City* (London: Routledge).

Skeggs, B. (2004), *Class, Self, Culture* (London: Routledge).

Thaden, L.M. (2004 [1938]) *High, Wide, and Frightened* (Fayetteville: University of Arkansas Press) (Foreword by Patty Wagstaff).

Urry, J. (2000), *Sociology beyond Societies. Mobilities for the Twenty-first Century* (London: Routledge).

Uteng, T.P., and Cresswell, T. (eds) (2008), *Gendered Mobilities* (London: Ashgate).

Verstraete, G and Cresswell, T. (eds) (2002), *Mobilizing Place, Placing Mobility: The Politics of Representation in a Globalized World* (Amsterdam: Rodopi).

Vowles, T.M. (2006), 'Geographic perspectives on air transportation,' *Professional Geographer* 58(1): 12–9.

Ware, S. (1993), Still *Missing. Amelia Earhart and the Search for Modern Feminism* (New York: W.W. Norton & Company).

Wohl, R. (2005), *The Spectacle of Flight: Aviation and the Western imagination, 1920–1950* (New Haven: Yale University Press).

Wylie, J. (2005), 'A single day's walking: Narrating self and landscape on the South West Coast Path,' *Transactions of the Institute of British Geographers* NS 30, 234–47.

PART II
Spaces

Chapter 7
Roads: Lawrence Halprin, Modern Dance and the American Freeway Landscape

Peter Merriman

In writings on mobility and non-representational theory, academics tracing the more-than-representational, performative, expressive improvisations of bodies-in-movement-in-spaces have adopted a range of theoretical and methodological approaches to explore the production and complex entwined performativities, materialities, mobilities and affects of *both* human embodied subjects *and* the spaces/places/landscapes/environments which are inhabited, traversed and perceived. Important work has emerged in geography on the materiality and embodied practices associated with cities (Amin and Thrift 2002; Latham and McCormack 2004; Pinder, this book), the representational and non-representational dimensions of architecture (Lees 2001; Adey 2008; Kraftl and Adey 2008), the emergence of self *and* landscape whilst walking (Wylie 2005, 2006, 2007), the mobilities, materialities and embodied practices associated with driving and other ways of moving (Sheller and Urry 2000; Cresswell 2006; Merriman 2007), and a whole host of other themes. While a few geographers have quite deliberately attempted to provide accounts of the production *and* consumption (or inhabitation) of particular landscapes, architectures or environments (e.g. Lees 2001; Llewellyn 2004; Merriman 2007; Kraftl and Adey 2008), the majority tend to focus *more* on *either* the physical production of architectures and environments *or* how they are inhabited and experienced; either examining the planning, design, and production of environments, or drawing upon participative and ethnographic methods and phenomenological and post-phenomenological philosophies to examine how embodied subjects sense, inhabit or apprehend their surroundings.[1] Research on driving and roads is a case in point. In the social sciences and humanities, scholars have tended to focus *either* on the consumption and inhabitation of the micro-spaces of the car and the generic practices of driving along roads, *or* the

---

1 For a number of reasons I do not like the distinction between producing and consuming landscapes as it suggests a one-way flow, permanence in physical form and it can divert attention away from the production of experiencing subjects. I am more interested, here, in attempts to co-produce environments and subjects – as, indeed, are many other post-structuralist geographers drawing upon a range of methods. However, the distinction between production and consumption suits my purposes here, as many academics do still operate with such distinctions.

design, construction and landscaping of specific driving environments. There are clearly practical advantages to such divisions of labour, but while academic disciplines and traditions of enquiry may maintain or reinforce such divisions, many architecture, landscaping and engineering professionals have worked hard to overcome such distinctions, and throughout the twentieth century they developed new approaches to both driving *and* driving environments: in an attempt to *comprehend* how drivers perceive, inhabit, and experience the landscapes of roads and *develop* more sophisticated design, engineering and landscaping principles (see Appleyard et al. 1964; Kemp 1986; Schwarzer 2004; Merriman 2006, 2007; Zeller 2007). In this chapter I examine the work of one such design professional, the prominent San Francisco landscape architect and environmental planner Lawrence Halprin, whose writings and commissions incorporated approaches to movement, embodiment and choreography derived from modern dance and avant-garde performance. Indeed, Halprin's work can be situated in a long history of attempts by architects, dance choreographers, musicians and geographers (such as Torsten Hägerstrand) who have attempted to choreograph, notate, codify and diagram embodied movements in the environment (Pred 1977; McCormack 2005, 2008; Cresswell 2006). In Halprin's case, he approached streets, roads and freeways not simply as inactive material environments in which driving or walking could be practised, but as active spaces which must be carefully designed, 'scored' and choreographed to produce particular movements, experiences, emotions and affects for motorists, pedestrians and local communities. In this chapter I explore how Halprin drew upon particular knowledges and understandings of movement and the environment in an attempt to engineer the affective potential of particular landscapes, including those of roads.

Lawrence Halprin, Architecture and Modern Dance

> ...I wish to point out my debt to my wife, the dancer Ann Halprin, for many years of stimulating exchanges of ideas on concepts about movement and its relation to our total environment as part of a long search for the meaning of art in our society. (Halprin 1963, 4)[2]

As Lawrence Halprin acknowledged in his 1963 book *Cities*, his approach to architecture, movement and creative exploration was strongly influenced by discussions and collaborations with his wife, the pioneering avant-garde dancer Anna Halprin (on Anna's work see A. Halprin 1995; Worth and Poyner 2004; Ross 2007). Lawrence and Anna had met in 1939 and married in 1940, and both had realised the possibilities of collective and cross-disciplinary collaboration when Lawrence studied landscape architecture, design and architecture at

---

2   References to 'Halprin' are to works by Lawrence Halprin. Publications by Anna Halprin are referenced in the text as 'A. Halprin'.

the Graduate School of Design at Harvard University between 1941 and 1943 (Halprin 1986; Ross 2007). At Harvard, Lawrence was taught by and socialised with leading Bauhaus *émigré* thinkers such as Walter Gropius, Marcel Breuer and Laszlo Noholy-Nagy, and the modernist landscape architect Christopher Tunnard, while Anna sat in on a variety of design and architecture lectures. The couple revelled in the comprehensive, interdisciplinary, communal artistic training, and the belief in artistic unification and workshop experimentation, that was promoted by Gropius in line with Bauhaus principles, and both began to see the *process* of creativity as more important than artistic *form* (Halprin 1969; Ross 2007; on Bauhaus educational principles see Bayer et al. 1975; Whitford 1984). Anna began to realise that architects and dancers were experimenting and creating in similar ways, and in a speech on 'Dance and architecture' given to the Harvard School of Design c.1943–4 'at the suggestion of Walter Gropius', she explained how dancers and architects were both concerned with exploring and controlling space:

> The only difference is that in the dance, the definition of space and the experience of space takes place at one time, because we are defining space with movement and we experience space by movement. The architect, on the other hand, must define space first with materials, and then that space becomes a living force only when people can move around in that space and experience it through their movements. (A. Halprin 1943, 2)

Dancers simultaneously perform the dance and the dance-space, while architects are seen to be more like theatre designers or choreographers. Lawrence Halprin began to explore the relationships between architecture, landscape, dance and performance in his own work (after a period of naval service towards the end of World War Two), by which time the couple had settled in San Francisco. In a 1949 article on 'The choreography of gardens' Lawrence explained how gardens must be 'designed with the moving person in mind' (32) in order to allow them to participate in the landscape's performance:

> It has been brought to my attention that my gardens are like stage sets for a dance in that they are designed to determine the movement of the people in them. This has been a very conscious effort on my part … If the kinesthetic sense … can be cultivated and encouraged in our daily lives in garden and house and all our environment by designing for constantly pleasant movement patterns, our lives can be given the continuous sense of dance. (Halprin 1949, 31, 34)

In a series of prominent public landscape designs of the 1950s and 1960s, Halprin explored people's kinaesthetic sensibilities towards and apprehension of a range of environments (Halprin 1986). In designs for outdoor shopping centres/malls at Skokie (Illinois, 1955), Oakbrook (1959), Minneapolis (1962), and a number of public spaces in San Francisco (including Ghirardelli Square, 1962), Halprin used planning, design and landscaping techniques to choreograph pedestrian

movements through his spaces (Halprin 1986, 120, 123, 126–7).[3] Such exercises
in architectural choreography clearly have political resonances, as architectural
environments can form key components of programmes and techniques of
government, discipline and control, but Halprin was driven by liberal democratic
impulses and a belief that cities must have 'great diversity' and allow 'for freedom
of choice' (Halprin 1963, 7). As he explained in *Cities*:

> What we are really searching for is a creative process, a constantly changing
> sequence where people are the generators, their creative activities are the aim,
> and the physical elements are the tools. (Halprin 1963, 7)

Cities must be approached as 'a complex series of events' (9). They must 'provide
for those random and unforeseen opportunities, those chance occurrences and
happenings which are so vital to be aware of – the strange and beautiful which
no fixed, preconceived order can produce' (9). At a time when Anna Halprin was
emerging as a pioneering practitioner of the avant-garde, situated performances
known as 'Happenings', Lawrence began to explore everyday, mundane *and* staged
performance events and happenings in particular architectural environments.[4]
These explorations developed into a series of joint month-long workshops entitled
'Experiments in Environment', which were organised by Anna and Lawrence
Halprin for architects and dancers from the mid-1960s (Burns 1967; Halprin
1986; Ross 2007; Merriman 2010), but Lawrence started to outline his conceptual
thinking on these matters in books such as *Cities* (Halprin 1963). Here, he outlined
the materials and processes by which architects and urban designers could shape
creative urban environments: providing open spaces; designing functional street
furniture, public sculpture, and using lighting and signage effectively; providing
well-designed flooring; controlling water; providing trees and plants; and most
importantly, understanding the city as 'an environment for choreography' which
'must be experienced through movement' (Halprin 1963, 193):

---

3   There are clearly parallels here with the approaches of architects, planners, and
landscape architects who have, for a long time, designed architectural environments and
spaces such as roads to be experienced in motion (see e.g. Crowe 1960; Cullen 1961;
Merriman 2007). Halprin himself hints at this genealogy, referring to Humphry Repton's
attempts to compose landscapes seen from roads (Halprin 1966, 29; Merriman 2006, 2007;
on Repton, see Daniels 1999).
4   There is an extensive literature on the avant-garde art-forms known as 'Happenings'
and 'Environments'. As Allan Kaprow, a pioneer of Happenings, has explained: 'The term
'environment' refers to an art form that fills an entire room (or outdoor space) surrounding
the visitor and consisting of any materials whatsoever, including lights, sounds and colour
…The term 'happening' refers to an art form related to theatre, in that it is performed in
a given time and space. Its structure and content are a logical extension of environments'
(Kaprow, cited in Henri 1974, 4; On Anna Halprin's pioneering 'Happenings', see
Kultermann 1971; Sandford 1995; Ross 2007).

... the whole city comes alive through movement as a total environment for the creative process of living. We call this chapter the choreography of the city because of its implication of movement and participation – movement of people, of cars, of flying kites, of clouds and pigeons, and even the change of seasons. (Halprin 1963, 9)

Building upon contemporary thinking in town planning, Halprin expressed a need to separate the automobile and pedestrian, providing effective and pleasurable circulation systems for different functions and speeds, but he was also forced to address a dilemma which is now at the forefront of debates in human geography. Architects and planners are trained to represent buildings and environments using plans, elevations and models, but how can they develop a language or technique for understanding, describing or diagramming particular activities, practices or movements (cf. McCormack 2005)? To rephrase Halprin's dilemma in terms of contemporary geographical debates: how can architects effectively describe, diagram and choreograph non-representational movements, actions and events? What kinds of 'movement scripts' are needed to 'better capture embodied practice' (Thrift 2000, 235)? Or, are our engagements with architectural environments simply 'unspeakable, unwriteable and, of course, unrepresentable' (Laurier and Philo 2006, 353; Thrift 2000)? Halprin presented the dilemma thus:

> It is true that any good designer or planner will *think*, while he is designing, of the activity that eventually will occur within his spaces. But he cannot design the movement, for he has no tools to do so. Even highway engineers who deal with movement have no method of describing it. (Halprin 1963, 208)

Whereas some contemporary geographical theorists might assert the futility of attempting to describe, understand and design such movements, events or happenings, Halprin was a landscape practitioner forced to address everyday, practical design problems, and he asserted the need to 'codify a transmittable or universally understandable system ... to program movement carefully and analyze it' (Halprin 1963, 209). Halprin subsequently developed a new notation system entitled 'motation' that would enable designers to work 'kinaesthetically', focusing 'primarily on movement, and only secondarily on the environment' (Halprin 1963, 209; also 1965, 1972).

Motation: Movement Notation and Environmental Experience

Halprin's motation system was inspired by notation systems previously developed in planning, dance, music and art. In architecture and planning, Donald Appleyard, Kevin Lynch and John Myer (1964) had developed techniques for notating visual sequences and experiences in their studies of *The View from the Road*, but Halprin claimed that *his* system focused primarily on 'movement' rather than 'the environment of movement', providing effective 'choreographic devices to register movement quality, character, speed, involvement with other mobile (or static) elements, and progressive spatial relationships including vertical and horizontal [movements and relations]' (Halprin 1966, 87).[5] Halprin preferred to align his 'motation' system with parallel systems developed in music by John Cage and Morton Subotnick, and in dance by Rudolf Laban, and he hoped that his new system would 'have universal application for every kind of movement':

> I assumed that such a system ought to be useful for designers working with pure movement: in dance and theater; for the newer choreographers whose aim has been to fuse sculpture and painting with theater; as well as for those of us designing for environment – architects, planners, and landscape architects. (Halprin 1965, 126)

The motation system was designed to be 'reasonably simple' and 'readable', comprising individual 'frames' in sequences demonstrating spatial and temporal progression (Halprin 1965, 128). The system utilised a standard paper form, enabling one to record or score movements by noting events in 'frames' on a 'horizontal track' (recording the location of objects and path of travel in plan form), a 'vertical track' (recording the position of objects as they appear across the visual horizon), and two small tracks recording 'distance' (which can be used to record rises and falls in elevation, as well as 'special events' such as 'sound, smell, colour, or rain'), and 'time' (which also enables one to indicate speed) (Halprin 1965, 129–30). Special symbols were used to notate architectural structures, landscape features, moving subjects/objects, directions of movement and view, and alignment and shape. Twenty-six basic symbols were developed, including symbols for humans, cars, different sizes of building, hills, bodies of water, trees, shrubs and fences (Figure 7.2). Dots were associated with moving things, with the single dot representing a human being. Cars were represented by a dot enclosed in a circle, clouds by a dot enclosed in a semi-circle, and dots also formed part of the symbols for 'running water' and 'fountain' (Halprin 1965). Halprin acknowledged that 'patience and practice' would be required to learn this new symbolic language, but once grasped, the notation system could be used for 'recording existing events involving mobility and for designing for mobility' (Halprin 1965, 130). Motation, then, was not simply a symbolic linguistic system for *describing* or

---

5   While Halprin (1966, 87) claims that Appleyard, Lynch and Myer were primarily concerned with the 'environment of movement', their approach was not that different to Halprin's. Their techniques for notating highway visual sequences developed existing architectural and planning techniques (particularly Philip Thiel's (1961) 'sequence-experience notation' system), although, like Halprin, they also draw parallels with Laban's techniques for notating movement in dance (i.e. Labanotation) (see Appleyard et al. 1964, 21–23; on Labanotation, see Thrift 2000; Cresswell 2006).

recording movements through environments, but it was also to serve as a tool for *choreographing* or *scoring* future movements in environments:

> Motation is a tool for choreography as much as description; choreography in the broadest sense – meaning design for movement. … Motation can be used for choreographing dances for stage and theatre, for the design of movement through urban spaces at pedestrian speeds, or for the qualities of motion through space at the speed of freeways and rapid transit systems. (Halprin 1965, 130)

Halprin's motation system formed an important element in his design work and collaborations with Anna in the 1960s, but it was not without its problems. Indeed, while Halprin claimed that his notation system focused on the qualities and character of *the movement*, many of his published notations focus largely on the visualities generated by particular embodied movements and actions, and we are provided with little sense of the multi-sensory, kinaesthetic dimensions of performing mobile embodied actions. This is particularly apparent in two scores for quite different movements which were notated on 12 July 1965 and were used to illustrate his 1965 article on Motation and his 1966 book *Freeways*. The first score is of a dance sequence undertaken by members of Anna's dance workshop on her dance deck in woods near the couple's house at the foot of Mount Tamalpais in Marin County (Figure 7.1).

Lawrence and modern-dance lighting designer Arch Lauterer had designed the deck in the early 1950s 'specifically for movement experience':

> The space itself is alive and kinetic – it is changeable – it invites movement – challenging it by its own sense of movement … There is great change on the deck. The light moves through the trees… The seasons change … Dance here is only one of many moving elements. (Halprin 1956, 23–24)[6]

The Motation score recorded the dance movements in the dance-space, but what is significant is that the notator (presumably Lawrence) was not performing the movement and was instead observing the dance from the audience area. Hence, Lawrence's engagement with the dance is a largely visual one, and he is limited to providing visual descriptions of the dance rather than charting the kinaesthetic

---

6   The deck was a key space in the development of Anna's experimental approach to both performing and choreographing environmental dance. This was partly because the deck was out doors, and partly due to its irregular shape and layout. Indeed, her explorations of dance in this space pre-empted later attempts by artists and curators to escape the static space of the gallery or 'white cube': 'The customary points of reference are gone and in place of a cubic space all confined by right angles with a front, back, sides and top – a box within which to move – the space explodes and becomes mobile. Movement within a moving space, I have found, is different than movement within a static cube' (A. Halprin 1956, 24).

Figure 7.1   Dance Motation for the San Francisco Dancers' Workshop by Lawrence Halprin. Reproduced by permission of Lawrence Halprin Collection, The Architectural Archives, University of Pennsylvania

sensations of performing the dance in the dance-space. In the second example – a record of a four-mile car journey in San Francisco along the Embarcadero Freeway, San Francisco Skyway and Central Freeway, finishing at the Civic Center – the notator was in movement, recording the driver's view of the journey (Figure 7.2).

The 'horizontal track' shows the plan-form of the road and visible structures, the 'vertical track' documents the structures which enter the driver's view, while the 'distance' and 'time' tracks chart changes in elevation and speed, respectively. And yet, while Lawrence was moving through the landscape, noting the car's movements, the score still focuses on the visualities and materialities of the

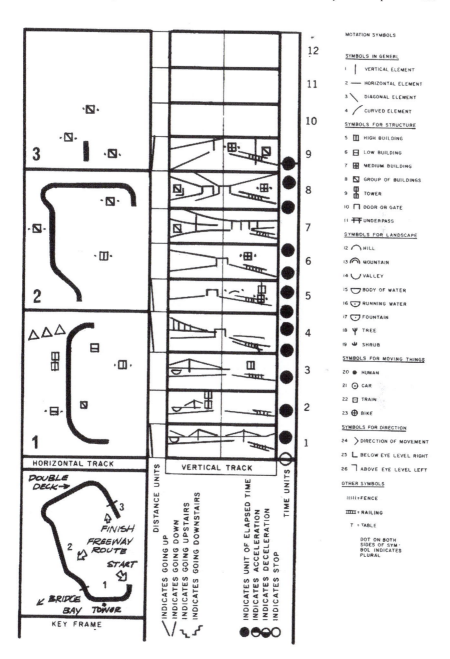

Figure 7.2   Freeway Motation by Lawrence Halprin. Reproduced by permission of Lawrence Halprin Collection, The Architectural Archives, University of Pennsylvania

freeway environment, and we are provided with little sense of the multi-sensory, kinaesthetic, and sensational dimensions of driving in this landscape.

In the late 1960s, Halprin attempted to develop a more nuanced approach to the creative process in architecture which could start to address some of the limitations of the motation system and the awkward relationship between the representational imperative expressed through 'scoring' and 'scores' and the more-than-representational aspects of all performances:

> Much of my own professional life has been involved in this apparent dichotomy: between the score and the performance ... I found that by themselves scores could not deal with the humanistic aspects of life situations including individual passions, wills, and values. (Halprin 1969, 1–2)

Halprin wanted to develop a more sophisticated understanding of the performance of particular events, aesthetics and values in architectural environments, and a more inclusive and participative approach that entailed planning *with* communities. This led him to develop the 'Take Part' participative design workshops (Halprin and Burns 1974) and the RSVP theory of the creative process: a way of schematising the creative process that suggested that there are four interrelated procedures and elements – Resources (R), Scores (S), Valuaction (V), Performance (P) – which must be visible and facilitated in a truly democratic, open, inclusive 'process-orientated society' (Halprin 1969, 2; Merriman 2010).[7]

Choreography, Freeways and City Streets

Lawrence Halprin's evolving attitudes to environmental design, choreography, urban aesthetics, and planning for communities are particularly evident in his writings and commissions relating to streets and freeways. As an urban designer and landscape architect Halprin was well-aware of the positive *and* negative impacts of the automobile and roads. In his 1966 book *Freeways* he describes the excitement and exhilaration of driving along beautifully designed sculptural freeways. The sensation is 'almost like flying' or 'swimming with fins' (1966,

---

7   The RSVP cycle is an ongoing process that can be executed in any order in multiple iterations, and it lay at the core of Halprin's inclusive approach to collective environmental design and community participation in the late 1960s and 1970s. Halprin's description of the four components of the cycle states that: 'R – *Resources*, which are what you have to work with. These include human and physical resources *and* their motivations and aims. S – *Scores,* which describe the process leading to the performance. V – *Valuaction,* which analyses the results of action and possible selectivity and decisions. The term 'valuaction' is one coined to suggest the action-oriented as well as the decision-oriented aspects of V in the cycle. P – *Performance,* which is the resultant of scores and is the 'style' of the process' (Halprin 1969, 2).

12), and these involving-structures should be counted 'among the most beautiful structures of our age':

> Freeways out in the countryside, with their graceful, sinuous, curvilinear patterns, are like great free-flowing paintings in which, through participation, the sensations of motion through space are experienced. In cities the great overhead concrete structures with their haunches tied to the ground and the vast flowing cantilevers rippling above the local streets stand like enormous sculptures marching through the architectonic caverns. These vast and beautiful works of engineering speak to us in the language of a new scale, a new attitude in which high-speed motion and the qualities of change are not mere abstract conceptions but a vital part of our everyday experiences. (Halprin 1966, 17)

Freeways and freeway traffic provide a vital movement that animates and enlivens the city, opening up 'new vistas', 'panoramic views' and providing 'some of the greatest new urban experiences' (1966, 23). Freeways provide new aesthetic experiences of the city; new impressions of 'color, form and shape' and 'a brilliant kaleidoscope of motion' (1966, 23). Halprin's impressionistic descriptions of the dynamic and, he suggests, 'filmic' visualities of urban freeway driving echo those made by other artists, architects and cultural commentators (see Merriman 2006, 2007). Drivers are enlivened and transported by the freeway's sculptural forms, while the movements of automobiles and drivers animate urban and rural landscapes. Movement is posited as vital to modern life, and the task of the architect, engineer, planner and landscape architect must be to choreograph the multifarious movements bringing life to these landscapes. With rural freeways, engineers and landscape architects must carefully choreograph the landscape, creating roads which 'give us the excitement of an environmental dance, where man can be in motion in his landscape theater' (Halprin 1966, 37). Urban freeways, on the other hand, should be treated as 'a new form of urban sculpture for motion' with 'inherent qualities as works of art' (Halprin 1966, 5).

Halprin's assertions and vision were quite emphatic: 'freeways, as well as every other transportation mechanism, must be an *integral* part of the rebuilding of cities' (Halprin 1966, 113). Freeways are, here, positioned as important elements in the future life of cities, but the key word is 'integral', for Halprin was also clear that the vast majority of freeways constructed up to that time were not *integral* to, or fully integrated into, cities. Urban freeways all-too-frequently rupture and disintegrate the fabric of urban communities, as highway designers adopt rational and abstract planning principles, overlooking the subjective, sociological and aesthetic aspects of planning, design and construction. Freeways *could* be works of art, engineering masterpieces, but many US freeways 'demeaned the cities they were meant to serve' (Halprin 1966, 5). Echoing the concerns expressed by an increasing number of urban thinkers, most notably Jane Jacobs in *The Death and Life of Great American Cities* (1961), Halprin argued that too many freeways were disrupting and dividing neighbourhoods; destroying immeasurable and unquantifiable urban qualities and values. Elevated and stacked freeways had a particularly major impact on communities and landscapes, and Halprin described San Francisco's Embarcadero Freeway as one of the worst (Halprin 1966, 74). This elevated, two-tier, stacked freeway separated the city's downtown districts from the waterfront, and during its construction widespread opposition was already being voiced. By the time the freeway opened in 1959, the city's Board of Supervisors had voted to cancel seven of the ten freeways proposed for the city (Carlsson n.d.)

Halprin was not involved in designing the Embarcadero Freeway, but he worked on, around and under it in a number of projects. The second of the two Motation studies discussed earlier recorded the experience of driving along the Embarcadero Freeway. In 1962, Halprin was commissioned to design the four-acre Embarcadero Plaza (now the M. Justin Herman Plaza) adjacent to the freeway, and in designing this public space Halprin designed and choreographed a landscape that would be able to counter the visual spectacle, force and noise of the elevated freeway. The resulting plaza was completed in 1972 and centred around a concrete, sculptural fountain by the French Canadian sculptor Armand Vaillancourt. The sculpture, entitled *Québec Libre!*, was designed to 'echo the force of the nearby freeway and to encourage people's movement through, under, and in it' (Halprin 1986, 127). The undesirable spectacle of the freeway and sound of moving traffic was echoed and offset by a landscape animated with the thunderous movement of water, utilising choreographic and sculptural devices which he had discussed at length in *Cities* (Halprin 1963, 134–5, 158–161).

By the early 1960s, Halprin was a locally, nationally and internationally acclaimed environmental designer and landscape practitioner, and in 1962 the San Francisco Board of Supervisors commissioned Halprin to conduct a study of the impact of freeways on the city. In 1963 the California Department of Highways appointed Halprin to prepare a plan for the proposed Panhandle Freeway linking the Golden Gate Bridge and the Central Freeway, but his report provided much more than a simple plan suggesting where to locate the road. Halprin drafted a series of fundamental 'creative principles' which must underpin urban freeway design. Urban freeways must 'follow the grid pattern ... fit into existing and projected land-use and topographic patterns in a city ... be condensed and concentrated ... integrated with the city ... and built as part of a total community development' (Halprin 1964, 10; 1966). At a time when freeway proposals were receiving widespread opposition in San Francisco, Halprin's plans proved highly controversial. Resistance to the proposals reached a peak in 1964, with the foundation of the Haight-Ashbury Neighborhood Council, the staging of prominent rallies, and the eventual rejection of the plans by the Board of Supervisors in October 1964 (Carlsson n.d.). Nevertheless, Halprin's planning principles gained widespread publicity, and in 1965 he was one of eight individuals appointed to act as advisors to the US Government's Federal Highway Administrator. The group included landscape architects, planners, consulting engineers, structural engineers and architects, and in their extensive report on *The Freeway in the*

City (1968), the team espoused many of the design principles which had been adopted and advocated earlier by Halprin (Urban Advisors 1968).[8] Their report described how the US Interstate highway construction programme was 'under heavy attack', how the freeway programme had 'ignored the soul of the city', and how, in the future, freeways 'must contribute to the total city environment' (Urban Advisors 1968, 10–11). Sixteen major recommendations were put forward to the government, covering such matters as compensation, integration with other forms of urban transport, comprehensive planning, and the planned use of 'the space beside, below, or above the freeway' (19). What's more, in the same year as Lawrence was developing novel approaches to design education, working with architecture and dance students in the month-long 'Experiments in Environment' workshops run with Anna, the Advisors recommended government subsidies for 'experimental educational programs in our schools of environmental design' (17). Interdisciplinary planning teams should be formed, drawing upon the expertise of sociologists, economists, ecologists and political scientists, as well as engineers, planners, architects and landscape architects. Experts should work with communities, allowing them to 'participate in regional planning activities', and to express their viewpoint on proposed locations for freeways (27 and 35):

> Groups of citizens must be involved in the complexity of the value weighing process. ... only with their involvement can the city become expressive of the needs and feelings of its citizens. (Urban Advisors 1968, 15)

During the late 1960s communities were mobilising against disruptive and destructive redevelopment projects and civil rights leaders were pushing for equal rights for minority groups, while liberal planners like Halprin were calling for more inclusive, participatory and democratic planning techniques.[9] Halprin was to most fully develop these participatory approaches in his own 'Take Part' workshops with communities in the early 1970s (see Halprin and Burns 1974), but *Freeways in the City* also articulated techniques for minimising the impact of freeways on communities which Halprin had been advocating since the early 1960s (Urban Advisors 1968). For the remainder of this section I want to explore

---

8   The Urban Advisors included: Michael Rapuano (Chairman, landscape architect and consulting engineer), Lawrence Halprin (landscape architect and urban designer), Thomas C. Kavanagh (consulting engineer), Harry R. Powell (structural engineer), Kevin Roche (architect), Matthew L. Rockwell (architect and planner), John O. Simonds (landscape architect and planner), and Marvin R. Springer (urban planning consultant) (see Urban Advisors 1968, 3). While the report quite clearly echoes some of the planning and design principles which Halprin outlined in his books and articles, both Halprin and the team of advisors were heavily influenced by existing writings on these matters.

9   Anna and Lawrence both undertook projects *with* community groups, and both undertook extensive work with the black community at a time of ethnic tension and discussion about racial integration (see Ross 2007).

how two of these principles – separating vehicular and pedestrian traffic, and using 'earth' and 'air' rights under and over freeways – were utilised by Halprin in two commissions.

The principle of separating traffic of different functions and speeds was not a new one. Architects, town planners and urban visionaries had realised the benefits of such strategies for hundreds of years (see Halprin 1963, 1966), but it was only in the mid-twentieth century that architects and town planners were forced to grapple with – and in many cases mitigate against – the negative impact of the motor car on urban street-spaces. One of the first schemes requiring Lawrence to deal with such issues was his 1962 commission to redesign Nicollet Avenue in Minneapolis, Minnesota. This was also one of the first spaces Halprin 'scored' using the motation system, choreographing 'people-movement through its spaces' (Halprin 1969, 71). Nicollet Avenue was the main shopping street in Minneapolis, but it was losing its vitality, assuming a 'non-descript' appearance, it had rather narrow pavements, and was dominated by motor vehicles (72). Halprin's response was to partially pedestrianise eight blocks of the street, limiting access to buses, pedestrians and service vehicles, widening the pavements, creating a serpentine curve to the road, planting trees, and incorporating a range of sculptural and graphical devices (Halprin 1986, 127; 1969). Halprin's aim – like that of a dance choreographer or theatre designer (or, indeed, a prison architect) – was to choreograph the movements of people through the creation of a distinctive urban environment:

> Street configurations are scores because they control not only the patterns and rhythms of people (and vehicles, of course) but also the course, even the nature, of events within a city. (Halprin 1969, 82)

Halprin was a landscape architect and environmental designer concerned with the engineering of communities, movements, emotions and affects, as well as providing functional and aesthetically pleasing landscapes (on the engineering of affect in cities, see Thrift 2004). Choreography and scoring techniques were central to his plans.

Halprin employed similar techniques for separating traffic and pedestrians in other projects, and this was possibly at its most spectacular in his award-winning Seattle Freeway Park (commissioned in 1970) (Halprin 1986, 137). The Seattle Park Commission had approached Halprin about the possibility of establishing a park adjacent to Interstate 5, to which he suggested that they could build over the freeway, unifying the two neighbourhoods on either side of the road. Halprin's proposals articulated principles he had carefully outlined and advocated in *Freeways*: separating pedestrians and motor vehicles, condensing development, utilising 'air rights', and masking the noise, visual intrusion and physical separation of the freeway (Halprin 1966). The completed park included monumental rectangular concrete pillars, steps, extensive vegetation, and waterfalls 'design[ed] to drown out traffic noises' (Halprin 1986, 137).[10] The landscape was scored in order to draw people through it in particular ways and along particular routes. The landscape was choreographed to generate particular movements, sensations, aesthetic experiences and emotions, and to mask the sights, sounds and vibrations of the freeway. Halprin's Seattle Freeway Park appears as a carefully choreographed performance-space; a jungle of concrete, vegetation and waterfalls to be encountered and traversed by people-in-movement. The park has been described as 'one of Halprin's most influential urban design projects', and it received prominent awards by the American Society of Landscape Architects, Association of Landscape Contractors of America, and *Design and Environment* magazine upon its completion in 1976 (Halprin 1986, 137).

Conclusion

> ... the new theatre-dance and the environment as Ann and I have been practicing them are nonstatic, very closely related in that they are process-oriented, rather than simply result-oriented. Both derive their strengths and fundaments from a deep involvement in activity ... Both deal with subtleties and nuance, intuition, and fantasy, and go to the root-source of human needs and desires – atavistic ones at that. (Halprin 1969, 1)

As Lawrence explains in the introduction to *The RSVP Cycles* (1969), he and Anna were both concerned with process, practices and movement in their separate and collaborative efforts to choreograph bodies-in-environments. Lawrence engaged with ideas developed in modern dance and music, as well as architecture, planning and landscape architecture, in an attempt to comprehend, score and choreograph dynamic actions and events in architectural environments. These performative embodied movements were highly contextual, dynamic and frequently ephemeral, generating emotions and affects which are difficult to represent or quantify, but Halprin was quite clear that as a designer it was his job to engage with and engineer people's embodied movements, sensations and experiences. Urban streets and freeways provided important spaces in which Halprin attempted to choreograph movements and actions, and in approaching these spaces he recognised that he was choreographing, designing or engineering both the spaces *and* the movements, the environments *and* the actions. Spaces and practices must be seen as intricately

---

10   Since it opened in 1976 the park has gained a reputation as a crime hotspot, with the occurrence of at least one murder, a number of rapes, and numerous muggings. While Halprin envisioned the park as beholding aesthetic surprises and utilising water to drown-out traffic noises, critics have highlighted how the thunderous traffic and water may have drowned out the screams of the murder victim and how the maze-like paths and low lighting hampered crime prevention (see Mudede 2002). For more detail on Lawrence's work on the park and its subsequent use, see Hirsch (2006).

intertwined, and as such it is perhaps more useful to speak of 'movement-spaces' than of either 'movement-in-spaces' or 'spaces for movement'.[11] To draw upon the language of contemporary human geography, Lawrence and Anna Halprin, like all artists, architects and engineers, were professional practitioners of more-than-representational arts, and they were clearly at times more-than-representational and non-representational theorists. Lawrence Halprin recognised the inherent, persistent and ongoing tension between the representational and more-than-representational/performative, and he was actively engaged in projects which were about attempting to generate and engineer particular sensations, emotions and affects. What's more, looking back at Halprin's work, at a time when critics of non-representational theory have asked what the political import or impact of this work is, it is clear that his writings and community design projects articulate both a more traditional radical sense of politics and an understanding of the politics of affect (and of engineering affect) (Thrift 2004; McCormack 2005). Halprin combined, then, a more traditional radical politics aimed at democratising planning and including minority groups in the decision-making process, with an awareness and engineering of the kinds of politics of affect, emotion and movement which are currently at the forefront of discussions in cultural geography.

Acknowledgements

I would like to thank: Anna Halprin and the San Francisco Performing Arts Library and Museum for permission to reproduce unpublished quotations from materials in the Anna Halprin Archives; William Whitaker at The Architectural Archives of the University of Pennsylvania for permission to reproduce images from the Lawrence Halprin Collection; and audiences of earlier versions of this paper/ material which were delivered at AAG Annual meetings in San Francisco 2007 and Boston 2008 and the AHRC funded workshop 'On the Go' at Royal Holloway, University of London in April 2008.

References

Adey, P. (2008), 'Airports, mobility and the calculative architecture of affective control', *Geoforum* 39, 438–51.
Amin, A. and Thrift, N. (2002), *Cities* (Cambridge: Polity).
Appleyard, D., Lynch K. and Myer, J.R. (1964), *The View from the Road* (Cambridge, MA: The MIT Press).
Bayer, H., Gropius, W. and Gropius, I. (eds) (1975 [1938]), *Bauhaus 1919–1928* (New York: The Museum of Modern Art).

---

11   I have borrowed the term 'movement-space' from the title of a paper by Nigel Thrift (2003).

Burns, J.T. (1967), 'Experiments in environment', *Progressive Architecture* July, 130–7.

Carlsson, C. (n.d.), 'The freeway revolt', *Shaping San Francisco Digital Library* (http://shapingsf.org/), accessed 29 March 2008.

Cresswell, T. (2006), *On the Move: Mobility in the Modern West* (London: Routledge).

Crowe, S. (1960), *The Landscape of Roads* (London: The Architectural Press).

Cullen, G. (1961), *Townscape* (London: The Architectural Press).

Daniels, S. (1999), *Humphry Repton: Landscape Gardening and the Geography of Georgian England* (London: Yale University Press).

Halprin, A. (1943), 'Dance and architecture: a speech given at Harvard for the Design School at the suggestion of Walter Gropius'. Unpublished paper in San Francisco Performing Arts Library and Museum (SFPALM), Anna Halprin papers 11/62.

Halprin, A. (1956), 'Dance deck in the woods: Use of the deck', *Impulse Dance Magazine*, 24–5.

Halprin, A. (1995), *Moving Towards Life: Five Decades of Transformational Dance* (ed. R. Kaplan) (London: Wesleyan University Press).

Halprin, L. (1949), 'The choreography of gardens', *Impulse Dance Magazine*, 30–4.

Halprin, L. (1956), 'Dance deck in the woods', *Impulse Dance Magazine*, 21–4.

Halprin, L. (1963), *Cities* (New York: Reinhold Publishing Corporation).

Halprin, L. (1964), 'Why I'm in favor of the Panhandle Freeway', *San Francisco Examiner*, 3 May, 8–10.

Halprin, L. (1965), 'Motation', *Progressive Architecture* 46 (Part 2), 126–33.

Halprin, L. (1966), *Freeways* (New York: Reinhold Publishing Corporation).

Halprin, L. (1969), *The RSVP Cycles: Creative Processes in the Human Environment* (New York: George Braziller).

Halprin, L. (1972), *Notebooks 1958–1971* (London: The MIT Press).

Halprin, L. (1986), *Lawrence Halprin: Changing Places* (ed. L. Creighton Neall) (San Francisco: San Francisco Museum of Modern Art).

Halprin, L. and Burns, J. (1974), *Taking Part: A Workshop Approach to Collective Creativity* (London: The MIT Press).

Henri, A. (1974), *Environments and Happenings* (London: Thames and Hudson).

Hirsch, A. (2006), 'Lawrence Halprin's public spaces: design, experience and recovery. Three case studies', *Studies in the History of Gardens and Designed Landscapes* 26, 1–98.

Jacobs, J. (1961 [1965]), *The Death and Life of Great American Cities* (Harmondsworth: Pelican).

Kemp, L.W. (1986), 'Aesthetes and engineers: the occupational ideology of highway design', *Technology and Culture* 27, 759–97.

Kraftl, P. and Adey, P. (2008), 'Architecture/affect/inhabitation: geographies of being-in buildings', *Annals of the Association of American Geographers* 98, 213–31.

Kultermann, U. (1971), *Art-Events and Happenings* (London: Mathews Miller Dunbar).

Latham, A. and McCormack, D. (2004), 'Moving cities: rethinking the materiality of urban geographies', *Progress in Human Geography* 28, 701–24.

Laurier, E. and Philo, C. (2006), 'Possible geographies: a passing encounter in a café', *Area* 38, 353–63.

Lees, L. (2001), 'Towards a critical geography of architecture: the case of an Ersatz colosseum', *Ecumene* 8, 51–86.

Llewellyn, M. (2004), '"Urban village" or "white house": envisioned spaces, experienced places, and everyday life at Kensal House, London', *Environment and Planning D: Society and Space* 22, 229–49.

McCormack, D. (2005), 'Diagramming practice and performance', *Environment and Planning D: Society and Space* 23, 119–47.

McCormack, D. (2008), 'Geographies for moving bodies: thinking, dancing, spaces', *Geography Compass* 2(6): 1822–36.

Merriman, P. (2006), '"A new look at the English landscape": landscape architecture, movement and the aesthetics of motorways in early post-war Britain', *Cultural Geographies* 13, 78–105.

Merriman, P. (2007), *Driving spaces* (Oxford: Wiley-Blackwell Publishing).

Merriman, P. (2010), 'Architecture/dance: choreographing and inhabiting spaces with Anna and Lawrence Halprin', *Cultural Geographies*, in press.

Mudede, C. (2002), 'Topography of terror', *The Stranger* 22–28 August (http://www.thestranger.com/seattle/Content?oid=11685), accessed 12 June 2008.

Pred, A. (1977), 'The choreography of existence: comments on Hagerstrand's time-geography and its usefulness', *Economic Geography* 53(2): 207–21.

Ross, J. (2007), *Anna Halprin: Experience as Dance* (London: University of California Press).

Sandford, M.R. (ed.) (1995), *Happenings and Other Acts* (London: Routledge).

Schwarzer, M. (2004), *Zoomscape: Architecture in Motion and Media* (New York: Princeton Architectural Press).

Sheller, M. and Urry, J. (2000), 'The city and the car', *International Journal of Urban and Regional Research* 24, 727–57.

Thiel, P. (1961), 'A sequence-experience notation for architectural and urban spaces', *Town Planning Review* 32(1): 33–52.

Thrift, N. (2000), 'Afterwords', *Environment and Planning D: Society and Space* 18, 213–55.

Thrift, N. (2003), 'Movement-space: the changing domain of thinking resulting from new kinds of spatial awareness', *Economy and Society* 33, 582–604.

Thrift, N. (2004), 'Intensities of feeling: towards a spatial politics of affect', *Geografiska Annaler B* 86, 57–78.

Urban Advisors to the Federal Highway Administrator (1968), *The Freeway in the City: Principles of Planning and Design. A Report to the Secretary, Department of Transportation.* (Washington, DC: US Government Printing House).

Whitford, F. (1984), *Bauhaus* (London: Thames and Hudson).

Worth, L. and Poyner, H. (2004), *Anna Halprin* (London: Routledge).

Wylie, J. (2005), 'A single day's walking: narrating self and landscape on the South West Coast Path', *Transactions of the Institute of British Geographers* 30, 234–47.

Wylie, J. (2006), 'Depths and folds: on landscape and the gazing subject', *Environment and Planning D: Society and Space* 24, 519–35.

Wylie, J. (2007), *Landscape* (London: Routledge).

Zeller, T. (2007), *Driving Germany: The Landscape of the German Autobahn, 1930–1970* (Oxford: Berghahn Books).

Chapter 8
Bridges: Different Conditions of Mobile Possibilities

Ulf Strohmayer

Introduction: Mobility and Geographies of the Built Environment

Of the many forms of technology conditioning mobilities in one way or another, buildings have not, for the longest time, been assumed to form individually determining factors. Where research focused on the built environment as a contributing factor in the initial construction and subsequent maintenance of mobilities, unitary and singular buildings were accorded a side interest at best, and were subsumed within larger infrastructures, networks, or relational effects created by an assemblage of buildings. And while recent work in cultural and social geography has begun to address the formative capacity of buildings, relevant areas of research such as transport planning and history continue to have little time for the individual building as such, while no less relevant fields of study such as art history or the engineering sciences do not normally concern themselves with the way socially relevant actions are linked to or indeed facilitated by buildings. The majority of geographical writings on mobility have followed this trajectory: unless researching the micro-geographies of households or the macro-geographies of spatial divisions of labour, architecture forms at best an assumed and largely stable set of geographical nodes into which – resembling the work of time-geographers a generation ago – mobilities are thrust. What had been absent until fairly recently is a sustained analysis of the conditioning work emanating from architectural practices – an analysis that is capable of reconciling recent insights into the centrality and fluidity of mobilities (Cresswell 2006; Sheller and Urry 2006) with theoretically based discussions about the nature and role of the built environment.

Recent work on mobility, loosely associated with what has been classified as a 'new mobilities' paradigm, has taken the necessary steps toward such a theoretically informed and empirically centered appraisal of the built environment. Peter Kraftl's work on individual houses (2009) and their performative capacities (2006) should be mentioned in this context, alongside the central invocation of buildings and their geographies in Loretta Lees' work on gentrification (2003), the iconography of material urban landscapes (2002) and her survey of the interface between architecture and geography more generally (2001). A shared interest in this work is the attempt to resurrect the materiality of buildings (and thus, by

implication, of the built environment) from an over and often exclusive focus on 'representation' that had come to dominate debates influenced by once 'new' forms of cultural geography.

Within this body of work, the one most closely concerned with mobility is probably the work on airports by Peter Adey (2008a and 2008b). In fact, irrespective of the connection to the theme of 'mobility', Adey's work has clear implications for any work at the interface between architecture and geography. His insistence on an irreducible materiality that forms a basis – qua architecture or the built environment – for any subsequent social action (mobile or not), is both theoretically astute and empirically grounded. Adey's work, furthermore, is informative in its attempt to position the built environment at the interface between determination and free will, or, between structure and agency. I shall return to this aspect of the built environment in the closing pages of this chapter.

The following pages aim to contribute to these and related debates on the relationship between mobility and architecture. It will do so through the use of a historically remote case study containing the potential to inform a theoretically infused inclusion of architecture into geographical writings on mobility.[1] More specifically, the chapter follows a line of reasoning set by Urry's brief discussion of the role of 'pavements and paths' (2007, 63–89) in the structuring of urban forms of mobility. Through the use of its particular case, the argument aims to add historical depth and breadth to Urry's analysis, which appears to reinforce the idea of the nineteenth century as the principal key to understanding contemporary forms of mobility. If anything, those sources to the understanding of our contemporary world are more varied and complex than has hithertofore been acknowledged.

History, Mobility and Architecture

A most immediate nexus between architecture and mobility comes in the form of bridges. Bridges often embody the rather stark and most certainly binary choice of mobility as being either possible or not: they enable the crossing of a river or a gorge where, in their absence, none or none as direct, ubiquitous and temporally stable, would exist (Harrison 1992). Historically, this functional difference emerges most directly where and when bridges collapse or burn down, thereby rendering their day-to-day function momentarily disrupted and thus felt most acutely. For this to register, we need only witness the centrality accorded to destroyed bridges in the annals of urban histories around the globe, which, we ought to recall, was a regular occurrence in days marked by the absence of civil engineers, architects and building regulations. The temporary absence of a bridge that had been providing a condition of possibility for local and trans-local forms of mobility often became

---

1   See King (1996), Lees (2001), Yacobi (2002), Llewellyn (2003) and Jenkins (2006) for contemporary attempts to introduce an explicit inclusion of architecture within geographical writings more broadly construed.

genuine, disruptive events in the urban history of a particular city. We can thus hardly conceive of a more directly facilitating technology as the one provided for by bridges. In a narrow, geographical sense, this is less banal than it may otherwise seem: entire cities were formed around relatively easy or strategically placed river crossings, thereby starting intra-urban patterns of settlement and inner-urban topographies that often display a sturdiness second to none in the canon of geographical facts. Historically-minded spatial scientists have repeatedly made reference to the longevity and structural impact of bridges, be they located in urban or rural locations (Harrison 1992; Rowley 1985).

But such descriptions of factual causalities are of limited interest beyond the confines of a narrowly conceived historical geography. If understanding is not merely geared towards linking formal aspects of the built environment with broader historical trends but seeks to understand more fully the myriad of complex and everyday enabling and limiting linkages between individual practices, collective routines and the built environment, we clearly ought to ask questions of a different kind. The difficulty of incorporating any part of the built environment into sustained considerations of mobility is thus to a large extent a difficulty tied in with the so eminently powerful, positively tangible quality of the built environment: the latter exists (or not) in no uncertain manner and is thus often read and analysed in a linear, congruent fashion (see Dennis 2008, 4–20). A bridge, however, and to return to the focus of the present chapter, is not a unified, homogenous structure. Nor do bridges function in a unilateral manner. As a result, the particular forms of mobility facilitated and enabled by their presence are historically specific and anchored in concrete existing contexts. Take, by way of contemporary examples, the difference between a highway or *Autobahn*-like bridge, the pedestrian crossing of same and an inner-city bridge that combines different forms of traffic: rather obviously, the structure of a bridge and the technologies used determine what forms of mobility are facilitated, and what forms are not. Sometimes such usages are directly regulated, as in the case of restrictions on the use of intra-urban motorways prohibiting their use by cyclists and – in the case of my current place of residence – 'L-'drivers or 'learners', while in other contexts restricted usage emerges from practical experience, design or the assessment of risks. Either way, a bridge is not automatically comparable with another bridge in terms of the forms of mobility facilitated; presence and absence, in other words, are not the only contributing factors – design, power, and historical context enter into the equation in no less direct a manner. In the present-day context marked by discussions attaching to the carbon-impact of different technologies, such differences are perhaps more obviously noteworthy than they may have been in bygone decades. But such legal or practical limitations and their consequences still take roots in a highly contemporary idea of what a bridge is all about – and how it contributes to the rather binary notion of mobility alluded to earlier on in this chapter. Historically, however, bridges formed part of a much wider condition of possibility for mobilities to emerge. It is to these that the chapter will now turn. In so doing, the chapter acknowledges recent work aiming to link the material presence of the built environment with a multiplicity of possible social actions; its stated historical focus, however, largely rules out a recourse to ethnographic methods (Normark 2006) or statistical data (Pooley, Turnbaull and Adams 2006) to access and subsequently interpret concrete practices, be they mobile or not.

Perhaps the prime example of a built environment that differs from our contemporary conceptualization of bridges and their functions were the built-over bridges that dominated many European cities until well into the eighteenth century. The 'ideal type' of a city dominated by built-over bridges was arguably the city of Paris, where all four bridges crossing the River Seine by way of the Île de la Cité had been constructed with houses for the better part of four centuries (Mislin 1978).

The primary reasons for the emergence of built-over bridges such as the Pont au Change were tied in with relative advantages offered by their location near to the Île de la Cité, the administrative heart of medieval and early modern Paris. These locational benefits were initially recognized by money changers and jewelers, the key advantage resulting from the fact that both activities (the changing of money and the fabrication and selling of jewelry) depended on the king's sustained good-will – proximity to his palace on the Île de la Cité mattered crucially. Just as important, however, was the vulnerability of the goods to theft en route to the place where commercial activities took place. It was hence preferable to establish a unity of purpose that materialized architecturally on the bridge. The resulting specialization of commercial activities on bridges, which often left a mark on their naming ('Pont-au-Changeurs', 'Pont-aux-Mauniers' in the Parisian context alluded to in this chapter; see Jung 1985, 95), is one key characteristic of many 'living bridges' from the beginning of the sixth century, when houses first appeared on the bridges of Paris (Boyer 1976, 160).

Thus, from the twelfth to the sixteenth century, a succession of bridges were built, collapsed repeatedly, re-built and re-designed, and increased in number – and at no point did the thought of constructing a bridge without houses materialize. Occasionally, a collapsed bridge even gave way to two bridges, as in the case of the collapsed Grand-Pont, which was reconstructed in 'dual built-over' fashion in 1296 as the Pont-aux-Changeurs and the Pont-aux-Meuniers (Jung 1985, 28–30). In fact, the construction of a bridge without houses in Paris had to await the first tentacles of Absolutism: the much-admired and still existing Pont Neuf, constructed at the turn of the seventeenth century. The everyday reality of a bridge not merely adorned by houses – as celebrated contemporary remnants of such structures as the Ponte Vecchio in Florence or less well-known spaces as the *Innere Bruecke* in Esslingen, Germany, suggest to many – but forming a genuine 'living bridge' (Murray and Stevens 1996) is hard to imagine today. It is best perhaps not to focus initially on the most obvious difference to the bridges we use today – the existence of houses – but to think of such bridges as genuine extensions of urban street-scapes. Entering such a bridge was often only felt by virtue of the tolls that existed until the seventeenth century and which formed a consistent bone of

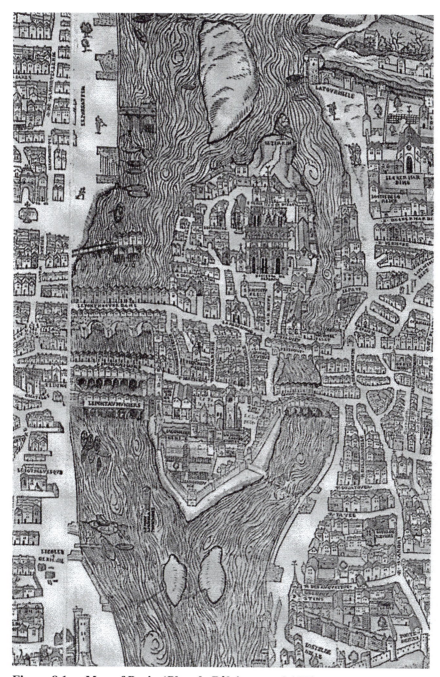

Figure 8.1   Map of Paris, 'Plan de Bâle', around 1550

contention in late medieval Paris (Boyer 1976, 99).[2] The street-scape of the bridge itself, however, did not differ from what preceded and followed it on urban spaces not built across water. As a sixteenth century traveller to the city of Lyon remarked upon encountering the city's built-over bridges:

> They [bridges] are erected with so much artifice that, being on these bridges, you are unable to judge whether you are on a bridge or streets continuing from the town, as much is the whole ornamented and clothed with magnificent houses. (Paradin 1573, 120)

Consequentially,

> [i]n the Middle Ages bridges were thoroughly integrated into town life. To reside on a bridge meant that one missed as little as possible of what was going on. (Boyer 1976, 160)

In marked similarity to the late medieval urban fabric, 'living bridges' were thus equipped with storage facilities, ateliers, shops, kitchens, latrines, sleeping rooms, as well as displaying a vertical social differentiation across the two or three floors initially constructed. In addition, such bridges often furnished mills underneath the arches of the bridges, toll and tax booths, as well as a (still unpaved) road left in the middle of the bridge to facilitate the crossing of the River Seine.

It is clearly the case that with such late medieval bridges, the infrastructure facilitating mobility was not exclusively – or even primarily – recognizable as such. In marked contrast to modern and often mono-functional roads or railroad tracks, these spaces could thus be seen to be ironically more akin to the post-modern spaces of consumption that have emerged at the nodes of mobility-supporting networks: the train stations or airports of the twenty-first century. The reason for this conflation of uses and the lack of a clear differentiation between 'mobile' and 'non-mobile' forms of urban existence, stem from historically specific sets of conditioning circumstances. Crucially, and somewhat in opposition to an emphasis on agency that permeates much writing on mobility today, these circumstances always potentially determine – and are in turn determined by – both the material context (in our case, the urban fabric of late medieval life) and the many agencies that are mobilized in their creation and maintenance. Agency here emerges as *structured* agency, conditioned by a historically contingent built environment that allows for possibilities to become actualities. We shall return to this problematic shortly.

---

2   For the most part, such tolls on the use of bridges were levied against non-local users only since a clear majority of local merchants, consumers and residents could apply for exemption from payments impeding upon their urban mobile practices.

The possibilities of mingling, of shopping, of producing and of living – in short, of being mobile and immobile simultaneously[3] – on say the built-over Pont-au-Change were thus markedly different from those that emerged on the quintessentially modern Pont Neuf, with its reduced range of functional diversity and its enhanced forms of mobility inducing technologies. On the Pont Neuf, a much wider street-scape, the introduction of pavement, the restrictions imposed on vending and other economic activities and, last but certainly not least, the original emphasis on the establishment of visual relationships across space all affected and continue to affect the very idea of mobility and its relationship to a set of enabling technologies. In fact, we could go as far as saying that the emergence of 'mobility' as a separate category of social existence – separate, indeed separable, from other socio-cultural forms of being – takes place in precisely such spaces, in the transition from built-over bridges to their modern, mono-functional counterparts. This 'carving out' of mobility from a more localized maelstrom characterizing social realities coincides with the move towards 'modern' forms of existence, with their more overt segregation of activities into discrete and classifiable categories. In this, 'mobility' is comparable with, and indeed embedded in, key aspects of social divisions of labour that emerge during the advent of modernity in many Western societies during the course of the last four hundred years.

It should be noted furthermore that the conditioning capacity of built-over bridges lead to a number of further, mobility-related and conditioning developments. Chief amongst these was the development of geographically fixed urban addresses or the sequential and sustained numbering of houses along a street (Bournon 1878): the first emergence of house numbers in an urban context can be documented on the Pont-au-Change in Paris in the context of the fifteenth century. It is quite possible that this novel form of imposing order onto the urban morphology was a result of the often striking uniformity of the houses constructed on these bridges, making the creation of a neutral and consistent 'address' desirable. The longevity and stability of the system created effectively bears witness to its practicable nature (Pronteau 1966; on the historical significance of numbering houses, see also Thale 2007 and Rose-Redwood 2008). But whatever the reason, the creation of such an administrative and postal target was to have clear implications for the social construction of mobile bodies, effectively contextualizing mobility both as a presence (as a permanent address) or as its other in the form of an absence, which from then onwards was increasingly used to define and codify both homeless and vagrant people. As such, this systematization and imposition of permanency onto urban landscapes was to become central to Enlightenment forms of urban governance, and since then as a condition of possibility for relating 'state' or 'commune' to individual citizens. It is difficult to imagine, by way of example, a census in the absence of a system permanently relating a particular body to a

---

3   Favier (1974, 15) notes that in fifteenth century Paris, 'mere' pedestrians opted to cross the Seine by using the much less stable Pont aux Meuniers rather than via the more centrally located Grand-Pont due the lack of 'flow' on the latter bridge.

clearly localized place.[4] Foucauldian biopower, the managing of people through the legislative regulation of bodily routines, once again takes roots within emergent geographical forms of governmental practices.

Planning Mobilities, Visualizing Space

What we encounter here, in a tentative conditioning capacity ascribed to the built environment, is increasingly embedded in normative contexts once the design of urban spaces becomes the domain of planners, urbanists and other professionally trained bureaucrats towards the end of the eighteenth, through the nineteenth, and more centrally in the first half of the twentieth century. As both Rabinow (1989) and Papayanis (2004) have convincingly demonstrated, the rise of 'planning' as both a profession and an analytical concern was centrally attached to the idea of a consciously executed design or 'plan' as a core tenet of modernity. We can now add 'mobility' as directly implicated – indeed, to a large extent invented – by that change as well. Needless to add that in all of these contexts, what has been alluded to as 'built environment' was and continues to be itself embedded within broader contexts such as the rise of capitalism, the development of state bureaucracies, changes in gender relationships, etc. Talk of a 'conditioning' power ascribed to structural environments and technologies therefore allows us to think the restructuring of environments to facilitate mobilities as a never neutral endeavour but as one that is deeply embedded in contexts which themselves require analysis. Contexts, we may add, that mattered then as much as they matter today. Let's take, by way of example, the construction of a motorway between two cities, where hithertofore a smaller road existed. Nevermind that bridges are an integral part of such construction activity. What concerns us here, is the fact that a new motorway will attract and create new traffic flows. In other words, it will enable new forms and patterns of geographical mobility. More than that, it will contribute to the solidification of car-centered forms of mobility, especially since its construction will more often than not come at the expense of alternatives not constructed. Once again, contemporary forms of mobility are clearly affected by planning decisions that change the structure of the built environment.

A key origin of such a planning-infused design of spaces that facilitate mobilities manifested itself in the construction of the Pont Neuf. As previously mentioned, this first new bridge originally constructed without houses in the Parisian context

---

4   Originally an administrative, rather than postal, practice, the subsequent numbering of houses in Paris was principally deployed to differentiate between 'authorised' and 'unauthorised' forms of development and thus quickly emerged as a means of differentiating between 'bourg' proper and the 'faubourgs' (Pronteau 1966, 73–79). The 'fixation' thus attaching to a house was resented by most as it also facilitated the implementation of new fiscal regimes – and had initially to be executed at night, as Sébastien Mercier observed in 1782 (Mercier 1782, 194, 'Les écriteaux des rues').

since the sixth century (Boyer 1976, 160) emerged within the context of what is one of the oldest and most openly ambitious building programmes in Western Europe (Ballon 1991). As such, the decision not to include houses in the design of the bridge – a decision to become paradigmatic in the context especially of an Enlightenment search for urban order and clarity[5] – was part and parcel of the invention of urban planning and thus a key element in the rise of modernity (see Strohmayer 2007). Interestingly, from a perspective that places mobility at the centre of its attention, consciously adopted changes to the built environment were thus directly implicating the very manner in which mobilities were technologically constructed and (or so we surmise) experienced. Bridges thus occupy a rather central place at the nexus between mobility, urban design and planning and mobility, leading to a set of mutually constituent moments in history.

In addition to such historical ruminations, we ought to note that the spaces of bridges have always been conceptualized and thought of differently from surrounding spaces, be they urban or rural. I am not initially thinking about an endemic romantic sentiment attaching to bridges in film, literature and the pictorial arts. More than most other places, perhaps, bridges symbolize a quite profound hybridity that is indeed characteristic of a particular type of space. Being both a site and a transition, bridges help us think mobility more clearly, perhaps to the point of being the architectural embodiment of mobility: one is here primarily because one is between places.[6] Bridges, to use the term employed by the German poet Ingeborg Bachmann in her poem 'Die Brücken' (*The Bridges*), are indeed quintessentially *einsam* (lonely) places that remain *namenlos* (nameless).[7]

> Besser ist's, im Auftrag der Ufer
> zu leben, von einem zum anderen,
> und tagsüber zu wachen,
> dass das Band der Berufene trennt.

> It's better to live at the river banks' bidding,
> from one side to the other,
> and to watch during days
> that the one who was called
> will sever the ribbon.[7]

---

5   A key commentator on the changes wrought of the urban fabric of Paris during the Enlightenment, the assistant of France's great eighteenth century architect Jacques-François Blondel, Pierre Patte expressly applauds the design of the 'uncluttered' Pont Neuf in his search for novel forms of urban expression (see Patte 1765, 9–11).

6   For a key contemporary analysis of how the in-between space of travel is filled, see Watts and Urry (2008).

7   My thanks to Eva Bourke for her translation of Bachmann's dense poem into the English language.

But then, the lesson learned from the Pont Neuf is that the in-between can afford its own perspective, can facilitate views unlike those made possible from other perspectives. In fact, following the work of Cosgrove and Daniels (1988), we may surmise that the consciously crafted and politically deployed openness of space on a non-built-over bridge like the Pont Neuf was instrumental in the creation of the Parisian urban landscape as we know it to this day: it was from the early seventeenth century onwards and directly related to the vistas facilitated by the 'open' bridge that engravings of Paris as a city by the River Seine, i.e. a city affording long views, began to proliferate.

The move away from the construction and maintenance of 'living bridges', which arguably was set in motion by the construction of the Pont Neuf, gained momentum throughout the eighteenth century until the dusk of the Bourbon reign coincided with the demolition of bridges across the French capital. A direct result of the Enlightenment obsession with clarity, and its translation into uncluttered, open spaces, urban bridges became what most of us associate with the term today: facilitators of mobility and of views.[8] As William Wordsworth observed as early as 1802 in his short sonnet 'Upon Westminster Bridge', bridges had indeed become structures that facilitated 'a sight so touching in its majesty' and thus capable even of 'naturalising' man-made structures like big cities. For poets and other human beings, a view from a bridge is unlike any other view, affording the individual not

Figure 8.2   Hubert Robert, *Demolition of the houses on the Pont au Change,* 1786-87, oil on canvas, Staatliche Kunsthalle, Karlsruhe

---

8   A key moment in this transition can be found in Pierre Patte's highly influential 1765 overview of Ancien Régime urban architecture, see especially 'Des Arts Liberaux – des Ponts et Chaussées', 9–11.

just with access to a fixed view of urban spaces but one that effortlessly becomes panoramic: the view of the city becomes itself a view in motion, sweeping in scope and cinematic in orientation, as Hart Crane astutely observed in his eminently mobile (and self-titled) 'P(r)oem: To Brooklyn Bridge' – itself both a prelude to his 1930 long poem 'The Bridge' and a reminiscence of Walt Whitman's poem 'Crossing Brooklyn Ferry' when the latter had been replaced by the former.

> [...]
> I think of cinemas, panoramic sleights
> With multitudes bent toward some flashing scene
> Never disclosed, but hastened to again,
> Foretold to other eyes on the same screen

Of course, the allusion to cinematic views serves as a reminder of commercialized and mass-produced pleasures in Crane's poem; the reference is here to the 'I think of ...' and its immediate link to visualized forms of communication emanating from the bridge.

Technologies, Speed and Conditioned Mobile Bodies

There is a final manner in which the Pont Neuf as a modern, house-less bridge is implicated in the workings of movement, one which attaches to architecture in general and to bridges specifically. In many geographically sensitive writings about mobility, it is the mobile body that is the focus of interest and rigour, thus perpetuating the humanist bias at the root of what has sometimes been called the 'new mobilities paradigm'. The fact that a lot of this research has adopted some form of 'agency' as its primary analytical centre of attention should not blind us to the realization that 'technologies' and other enabling 'structures' can be mobile even where and when the speeds involved are of a different kind than those shaping mobile bodies. Indeed, where speeds are congruent, geographers have little difficulty incorporating 'technological structures' into their theoretical frameworks – the train carrying the hobo or the camera registering movements in space are both in synch with the mobilities we observe (Cresswell 2001); but what about technologies or structures that are not?

Bruno Latour highlights this aspect when he analyses the Pont Neuf as an aging structure in need of repair, or replacement parts. It is this constant movement of parts, which is both operating at a different speed than the mobilities it facilitates and other than the normally assumed static existence of the built environment, that interests us here because it contains the possibility of a structurally cognizant and differentiated conceptualization of bodies, buildings and space. Latour writes:

> Yes, the Pont-Neuf, seized in its movement of renewed stone by stone, is part
> of public law. The difference between stone bridges, flesh and blood organs and

political bodies stems not from their nature but only from the pace at which their offices are renewed. (Latour 2004, n.p.)

Latour's reference to pace, in addition to his insistence on the importance of blurring the divide between nominally 'living' and 'mute' materials (as well as the difference between 'actors' and 'actants') which has proven to be inspirational in the context of Actor-Network-Theories, opens the way towards considerably less anthropocentric interpretations of 'mobility'. More important, however, than such a potentially toothless insight is his insistence on the structural bedrock of mobility – including the institutions that support, enable and prohibit mobilities, broadly construed. In the case of the Pont Neuf (effectively representing 'bridges' as a locale *sui generis*) these include – and include, crucially, in an original sense – a set of circulating bureaucracies, technologies, councillors and other 'institutions' in the sense attributed to the word by Latour. Here, then, we encounter another sense of mobility, one that is at the same time ubiquitous (and thus banal) and inescapable (and thus profound): the mobility that is the material world where things circulate and are constantly renewed, recycled, and replaced – and where, as a result, different speeds of mobility coexist, interact and conflict simultaneously (Guillaume 2007). Tracing the outer contours of these mobilities is what geographers do; since the flows of these mobilities are part of everyday routines that are thoroughly embedded within a global capitalist machine, they are best observed where aging things are in need of repair. Latour picked up on this notion when writing about the Pont Neuf, as did Léos Carax when he populated the originally unpopulated bridge with a group of homeless people in his *Les Amants du Pont Neuf* in 1989, shot during the closure of the Pont Neuf to traffic during extensive renovations in the run-up to the bi-centennial celebrations of the French Revolution (see Strohmayer 2002).

In this, Latour follows a path of theoretical interest charted some time ago by Walter Benjamin. It was Benjamin who insisted that the material world could best be analysed when it entered a state of decay, that it would reveal its innermost secrets only once it had become an immobile 'ruin'. Hence his life-long fascination with the Parisian Arcades that were to become a metaphorical *nomer* to Benjamin's unfinished project of a 'natural history' of the nineteenth century (Benjamin 2002; Buck-Morss 1989). We ought to remember that by the time Benjamin wrote about them, many Parisian Arcades – themselves hugely influential mobility-inducing and – changing structures – had become effective time-warps, spaces left behind by economic forces and cultural change. Not so the Pont Neuf, which continues to be central to Parisian everyday life, both in an economic and an aesthetic sense. Hence the even more pronounced need to attune to states of repair to see the involvement of different temporal regimes in the creation of mobilities that attach to both bodies and structures. The resulting fluidity of both, agencies and structures, is perhaps the most obvious of insights to warrant further studies.

It is this interface, the space between agency and determination that becomes apparent when the built environment decays or ceases to function much like a bridge that has collapsed, that allows us finally to tackle head-on the lingering notion of technological determinism that has been dormant in this chapter all along. A central issue that any sustained engagement with the built environment will encounter before long, the question of structured determination, broadly conceived, has always been key to understanding linkages between humankind and its many environments. And yet, notwithstanding its centrality, the question has been addressed in implied, rather than explicit, forms in the majority of geographical writings on the built environment. It is in this context that I propose to return to Peter Adey's formidable analysis of a viewing balcony at Liverpool Airport (2008a), for it is in this paper that readers can encounter one of the more direct engagements with the question posed. Abstracting for the moment from the particulars of his investigation, Adey's choice of words to approximate the role of architecture in the formation of new activities is highly instructive. When discussing the importance of the built environment for actual experience – the 'structural' contribution to agency so to speak – architecture is 'constructive', it 'ingain[s] specific beliefs', it 'inculcate[s]' and 'nurture[s] specific inspirations and ideas', or 'manufacture[s] a new way of being', as a not-quite random assemblage of adjectives and verbs on page 30 of the paper reveals. Further down, 'architectural geographies [...] worked to conduct feelings' (42) before the concluding part of the paper discusses 'how the airport could be sensed and felt in ways intended and designed for' (44). I have deliberately singled out this particular paper and its choice of language to highlight both the enormity of the task and a possible way forward: little less and nothing else than a subtle, situated approximation of the kind attempted by Adey will help us in our attempts causally to link the built environment and specific actions, practices and routines. At the same time, Adey's avoidance of terms usually associated with the structuring properties of structures, chief amongst which is the word 'determination' in its many shapes and guises, renders any gain emanating from such subtle approximations difficult to place within the larger landscape of learning.

Within the present chapter, lacking the direct recourse to individual and group-based sensory forms of experience that motivates so much of Adey's analysis due to the absence of voices or other forms of historical evidence speaking to life on pre- or early modern bridges, any link between an enabling condition of possibility in the form of a building (or parts thereof) and agency can ultimately only be speculative. At the same time, it is crucial to recognize that neither building nor experience is ever 'blank' (Adey 2008a, 42) but always already conditioned in turn by other buildings, other experiences and other contributing technologies. Crucially furthermore, as Adey has argued convincingly, a building may condition experiences most where it itself becomes 'invisible' or 'disappears', much like the Pont Neuf itself became a structurally hidden condition of possibility in the process of inventing the urban landscape. To me, perhaps parting company with Merriman's recent invitation to contemplate a more 'affective' or 'sensory' engagement in studies of mobility (2007), such analyses based on conditioning properties of architecture are no less desirable and perhaps more fruitful even, given the simple fact that collectively we can change the latter but have little way of influencing the former. What is more, as Latour, Benjamin and others have argued relentlessly, 'change' as concerns the built environment is inevitable, if only because *things age*, disrupting existing practices like mobility.

Conclusion

It is arguably the disruptive potential in the functioning of modern bridges that has turned them into metaphors centrally implicated in the landscapes of artistic expression. If the physical destruction of a medieval bridge could genuinely wreak havoc with local economies, the complete or temporal functional disruption of bridges in the more mediated worlds of today have long since acquired a metaphorical quality quite unlike those associated with other buildings or urban structures. By way of example, readers are invited to contemplate the use of a war-torn bridge as the primary stage for West Germany's first post-war key cinematic moment, Bernhard Wicki's *Die Brücke* (1959). Here a bridge becomes the main protagonist in a film marking the final days of the Third Reich through the utterly vain sacrifice of a couple of children in the defense of a strategically marginal bridge. Similarly, and with uncanny prescience, Ivo Andric's nobel-prize winning novel *Bridge on the Drina* (1945), employs a bridge to depict the changing fortunes of a town (Visegrad, Bosnia-Herzegovina, then Yugoslavia) at the border between Christian and Muslim Europe – prefiguring to some extent the highly symbolic destruction of the old bridge crossing the Neretva in Mostar in the context of the dis-unification of Yugoslavia in 1993.

Returning to the real worlds of today, it is somewhat fascinating to witness a development that places bridges yet again at the nadir of urban developments. For one, note the recently renewed interest in the relative structural messiness that accompanies 'living bridges'. From London to Hamburg, from Paris to small cities bordering rivers, urbanists have witnessed proposals for such quintessentially medieval bridges to be constructed at pivotal points within the urban fabric of the postmodern city. None have materialized to date. What has materialized, however, is a number of no less postmodern, playful allusions to a truncated functional capacity of bridges, now incorporated into the design of public buildings. Both Paul Chemetov's 1989 massive design for a new French Ministry of Finances (Ministère de l'Economie, des Finances et du Budget) in the Parisian twelfth arrondissement (see paris-architecture.info/PA-076.htm) and James Polshek's 2004 rather smaller building for the William J Clinton Presidential Center in Little Rock, Arkansas (see www.clintonpresidentialcenter.org), pay homage to nearby subway or disused railway bridges and to the genius loci by seeming to reach across river space, only to fail after the first arch. Here a functional diversity once characteristic of spaces that facilitated mobility beget mere architectural pointers towards a mobility that has long since become anchored in virtual spaces, rather than relying on a built environment to facilitate the movement of bodies across space. While the latter persist to provide important linkages for local economies, bridges today function on a host of scales and within a range of contexts, ranging from 'the real' to 'the imaginary' and beyond. They may not have lost their allure but certainly are no longer as centrally implicated in the construction of post-modern mobilities.

References

Adey, P. (2008a) 'Architectural geographies of the airport balcony: Mobility, sensation and the theatre of flight', *Geografiska Annaler, Series B: Human Geography*, 90, 29–47.

Adey, P. (2008b), 'Airports, mobility and calculative architecture of affective control', *Geoforum* 39, 438–51.

Ballon, H. (1991), *The Paris of Henri IV. Architecture and Urbanism* (Cambridge, MA: MIT Press).

Benjamin, W. (2002), 'Paris, Capital of the Nineteenth Century', in *The Arcades Project* (Cambridge, MA: Belknap).

Bournon, F. (1878), 'Le numérotage des maisons de Paris au Moyen Age,' *Bulletin de la societé de l'historie de Paris et de l'Île-de-France* 5, 138–40.

Boyer, M. (1976), *Medieval French Bridges: A History* (Cambridge, MA: The Medieval Academy of America).

Buck-Morss, S. (1989), *The Dialectics of Seeing: Walter Benjamin and the Arcades Project* (Cambridge, MA: MIT Press).

Cosgrove, D. and Daniels, S. (eds) (1988), *The Iconography of Landscape* (Cambridge: Cambridge University Press).

Cresswell, T. (2001), *The Tramp in America* (London: Reaktion).

Cresswell, T. (2006), *On the Move: Mobility in the Modern Western World* (New York: Routledge).

Dennis, R. (2008), *Cities in Modernity. Representations and Productions of Metropolitan Space, 1840–1930* (Cambridge: Cambridge University Press).

Favier, J. (1974), *Paris au XVe siècle, 1380–1500* (Nouvelle Histoire de Paris) (Paris: Diffusion Hachette).

Guillaume, V. (2007), 'Geographical airs – and areas' in Airs de Paris, exhibition at the Centre Georges Pompidou, Paris, 25 April–15 August 2007, available at www.centrepompidou.fr/PDF/AirsdeParis_ValerieGuillaume_en.pdf (accessed 2 August 2008).

Harrison, D. (1992), 'Bridges and economic development, 1300–1800', *Economic History Review* 45, 240–61.

Haw, R. (2005), *The Brooklyn Bridge: A Cultural History* (New Brunswick, NJ: Rutgers University Press).

Jenkins, L. (2006), 'Utopianism and urban change in Perreymond's plans for the rebuilding of Paris', *Journal of Historical Geography* 32, 336–51.

Jung, C. (1985), *Les Ponts de Paris à la fin du Moyen-Age* (Mémoire de Maîtrise) (Paris: Université de Paris IV).

King, R. (1996), *Emancipating Space: Geography, Architecture, and Urban Design* (London: Guilford Press).

Kraftl, P. (2006), 'Ecological architecture as performed art: Nant-y-Cwm Steiner school, Pembrokeshire', *Social and Cultural Geography* 7, 927–48.

Kraftl, P. (2009), 'Living in an artwork: the extraordinary geographies of the Hunterwasser-Haus in Vienna', *Cultural Geographies* 16, 111–34.

Latour, B. (2004), 'Plan 52, Step 13: Instituting', in *Paris: Invisible City*, available at www.bruno-latour.fr/virtual/EN/index.html (accessed 1 August 2008).

Lees, L. (2001), 'Towards a critical geography of architecture: The case of an Ersatz Colosseum', *Ecumene* 8, 51–88.

Lees, L. (2002), 'Rematerialising geography: the 'new' urban geography', *Progress in Human Geography* 26, 101–12.

Lees, L. (2003), 'Super-gentrification: the case of Brooklyn heights, New York City', *Urban Studies* 40, 2487–2509.

Llewellyn, M. (2003), 'Polyvocalism and the public: 'Doing' a critical historical geography of architecture', *Area* 35, 264–70.

Mercier, S. (1782), *Tableau de Paris*, Volume 2 (Paris).

Merriman, P. (2007), *Driving Spaces* (Oxford: Wiley-Blackwell).

Merriman, P. et al. (2008), 'Landscape, mobility, practice', *Social and Cultural Geography*, 9, 191–212.

Mislin, M. (1978), *Die überbauten Brücken von Paris* (PhD thesis) (Stuttgart: Technische Hochschule)

Murray, P. and Stevens, M. (eds) (1996), *Living Bridges: The Inhabited Bridge, Past, Present and Future* (Munich: Prestel).

Normark, D. (2006) 'Tending to mobility: intensities of staying at the petrol station', *Environment and Planning A* 38, 241–52.

Papayanis, N. (2004), *Planning Before Haussmann* (Baltimore: Johns Hopkins University Press).

Paradin, G. (1573), *Mémoires de l'histoire de Lyon* (Marseille: Lafitte).

Patte, P. (1765), *Mounuments érigés en France à la gloire de Louis XV* (Paris: Rozet).

Pierce, P. (2001), *Old London Bridge: The Story of the Longest Inhabited Bridge in Europe* (London: Headline).

Pooley, C., Turnbull, J. and Adams, M. (2006) 'The impact of new transport technologies on intraurban mobility: aview from the past', *Environment and Planning A* 38, 253–67.

Pronteau, J. (1966), *Les numérotages des maisons de Paris du XVI siècle à nos jours* (Paris: Couvert).

Rabinow, P. (1989), *French Modern: Norms and Forms of the Social Environment* (Cambridge, MA: MIT Press).

Rose-Redwood, R. (2008), 'Indexing the great ledger of the community: urban house numbering, city directories, and the production of spatial legibility', *Journal of Historical Geography* 34, 286–310.

Rowley, G. (1985), 'Of time, space and a bridge: central places in North-West Wales, c.1795–c.1860,' *Cambria* 12(1): 73–96.

Sheller, M. and Urry, J. (2006), 'The new mobilities paradigm', *Environment and Planning A*, 38, 207–26.

Strohmayer, U. (2002), 'Practicing Film: the Autonomy of Images in *Les Amants du Pont-Neuf*, in *Engaging Film: Anti-Essentialism, Space and Identity*, ed. D. Dixon and T. Cresswell (London: Rowan and Littlefield), 193–208.

Strohmayer, U. (2007), 'Engineering Vision: the Pont-Neuf in Paris and Modernity.' in *The City and the Senses: Urban Culture since 1500*, ed. A. Cowan and J. Steward (Aldershot: Ashgate), 75–92.

Thale, C. (2007), 'Changing addresses: social conflict, civic culture, and the politics of house numbering reform in Milwaukee, 1913–1931', *Journal of Historical Geography* 33, 125–43.

Urry, J. (2007), *Mobilities* (Cambridge: Polity Press).

Watts, L. and Urry, J. (2008), 'Moving methods, traveling times', *Environment and Planning D: Society and Space* 26, 860–74.

Yacobi, H. (2002), 'The architecture of ethnic logic: Exploring the meaning of the built environment in the 'mixed city' of Lop, Israel', *Geografiska Annaler, Series B: Human Geography*, 84(3–4): 171–87.

Chapter 9
Airports: Terminal/Vector

Peter Adey

Figure 9.1   Liverpool's Aerial Vectors (photo taken by author)

[T]he world is a flux of vectors, vectorial connections actualised in the events through which it pluralises itself by expressing its own energetic activity in variable configurations. (Alliez cited in Thrift 2006)

A word on this word 'vector' […] Virilio employs it to mean any trajectory along which bodies, information, or warheads can potentially pass […] vectors have fixed properties, like the length of a line in the geometric concept of vector. Yet that vector has no necessary position: it can link almost any points together. (Wark 1994, 11)

Introduction

'Where do those tubes go?' asks the character Lincoln Six Echo from Michael Bay's 2005 movie *The Island*. The tubes sitting in front of Lincoln (played by Ewan McGregor) are transparent tubes carrying red and blue liquids. Lincoln and his friend add some sort of chemical to the tubes as they fulfil their employment in the reality of a post-apocalyptic world. The protagonists have been told that they survived a disaster which has left the outside world 'contaminated', leaving the remaining inhabitants detained (unknowingly) in a containment facility controlled and managed by security personnel.[1] What keeps the members of the population motivated and apparently passive is the chance of going to 'The Island', a paradise departed for if one wins the weekly lottery. They are, however, destined for a rather different end. We soon learn that the characters' reality is quite unlike the one they are told. There has been no contamination. There is no 'island'. They are in fact clones, being grown and developed for eventual organ harvesting should their wealthy counterparts in the outside world need them.

I begin with this scene because Lincoln Six Echo's emergent and critical awareness of himself and his environment is enwrapped in a moment: when Lincoln asks, 'and where do those tubes go'. The characters merely see part of the tubes which pass through the laboratory/factory where they work on a daily basis. They see a portion, or a point of a vector of movement passing through their city. When the camera pans and follows the tubes out of the room and through the inner fabric of the building, the true secret of their reality is revealed. The audience sees that they are actually adding drugs and chemicals to an array of umbilical cords. The camera follows these cords into a room and into the bodies of thousands of clones or 'agnates' being grown in plastic bags. Lincoln and his friend are adding to the very lines of movement – mobilities of nutrients, proteins and other chemicals – of which they themselves are the 'product'. As these vectors point to the actuality of Lincoln Six Echo's situation, I think they do the same to our own.

These sorts of vectors are not just a figment of the imagination and they are often, as in Lincoln Six Echo's case, quite closed. I can look out of my window overlooking the city where I live and might see five or six contrails left by aircraft passing overhead (above). I look down onto the flyover nearby and see cars passing, making a wash in the rain-soaked tarmac. More permanent vectors exist too. My muddy footprints. The trail of a dripping tea-bag to the bin that I have failed to wipe up. Disturbingly, lines of dried blood on the pavement. Skid tracks in the car park across the road. Like the contrails in the sky, all of these lines denote their own kind of vector; a directional trace of a movement that once was. But vectors are not that easy to find unless one knows what one is looking for, nor are they just a remnant of the past. Vectors are 'of the now', and our awareness of vectors is heavily dependent upon how we are placed in relation to them. Lincoln Six Echo was caught somewhere in the middle, somewhere alongside the vector and thus unable to see its totality.

In a way this focus upon lines, paths and vectors sounds quite Hägerstrandian, and for good reason too. Torsten Hägerstrand was well aware of how his time-space routine diagrams – the vectors which he drew – represented the very 'tip' of experience. They had depth. He understood that, 'in the persistent present stands

---

1   What the 'disaster' actually was is never made clear.

a living body subject, endowed with memories, feelings, knowledge, imagination and goals [...] decisive for the direction of paths' (cited in Gregory 1979, 324). But while Hägerstrand famously argued that 'People are not paths, they just cannot avoid drawing them' (cited in Gregory 1979, 324), this is where our paths diverge, because I think that these paths or vectors are much more than the superficial 'tip' of mobility.

In this paper it is not my intention to reproduce the diagrammatic schemata of time-space routines, the networks and circuit models beloved by transport geographers (Haggett 1965; Lowe and Moryadas 1976), or the paths, points, nodes and lines, even of desire (recently re-invoked by Tiessen 2007) from spatial science. At the same time I don't want to turn away from them either. Much of the recent turn towards mobility has concerned itself with the spaces and paths of travel – from motorways to the lines of aeromobilities (Merriman 2007; Cwerner et al. 2008). Instead, I want to understand a dimension of people, things, and particularly airports as, quite literally, path-like. While Hägerstrand and more recent thinkers such as Tim Cresswell (2006) have understood these kinds of paths as abstractions, I suggest that vectors are embodied and experiential.

My examples come from various moments along the line of an airport. Interpretations of airports have been somewhat duplicitous in their thinking. As sites and infrastructures of vectors *par excellence* (Pascoe 2001), airports are also interpreted as considerably sunk and fixed entities (Adey 2006) or nodes (Crang 2002) through which masses and masses of things pass by way of the airport's facilitation of flight. But they are more than that. Being more-than representations of networks of lines and vectors, airports are made up of vectors – they are the lines of mobility that pass through them and that make them. The rest of this chapter untangles these lines through several different moments of Liverpool Airport, before reflecting on the power of vectors for the study of mobility in the conclusion.

Control

Vectors can be interpreted as representations that play their part in systems of monitoring and control. In this dimension, vectors are the means of pinning down mobility to a line. As already mentioned, in the past geographers saw the line of the network diagram as a way of giving order to the complex movements researchers tried to model, from the movement of people, livestock, grain, ideas and more (e.g. Hägerstrand 1967; Ullman 1957). In the contemporary airport, vectors of aircraft are representations used to manage and control the airport and air-traffic-control system. These vectors do not exist. They do not capture the movement of the aircraft but they describe where it is going and where it has gone. Resembling, in some way the lines drawn by Marey's and Muybridge's (Solnit 2003) time-motion studies, vectors strip off the bulk and the detail of movement in order to leave a sort of essence far easier to manage than the reality itself.

Modern airports are designed, run and managed along the notion of the vector and not just for air-traffic (Wells and Young 2004). Vectors give simplicity to the complexity of the airport. The airport terminal is commonly understood as a complex system of multiple users and stakeholders with communications and relationships between them (Graham 2003). In the planning of airports, flow-chart like diagrams are used to abstract and represent the possible movements of passengers through the various spaces and phases of airport experience, opening these mobilities up to the rationality of scientific abstraction and quantification. An interview with Liverpool Airport's architect was illustrated by lines, arrows and potential vectors of movement allowing us to emplace ourselves *at* the airport and *on* the airport's movements. Such paths denote both spatial mobility and the movement through the processes of the airport machine – from check-in to boarding (De Neufville and Odoni 2003). Passenger mobilities are treated indivisibly. They are imagined as flows and rivers and, thus, modelled as vectors that eventually become real in the 'real' material environment of the terminal. Lines and flows materialise into the tube like structures of gates, tunnels and corridors – the materialisation of what Deleuze and Guattari (1988) would know as 'hydraulic science'.

Vectors are used to make sense of and track potential 'risky objects' such as the explosives that could threaten the safety and security of passengers. 'Threat Vectors' define the paths of movement along which certain people, baggage and equipment may board a plane through an airport (Adey 2004). Because of the high numbers of vectors that run-through terminal spaces, addressing these paths is seen as 'a complex systems problem'. Putting people and objects under surveillance and scrutiny by adding more stringent controls as to who or what may pass, will supposedly reduce the chance of a threat making its way to the plane. In these terms, the vector absolves passenger intentionality and even their capacities for agency, but rather views people and spaces as paths and channels through which threats can pass, conduits along which objects can be transferred to their intended destination. Set in the context of a threat such as SARS, airports have witnessed how vectors mediate – allowing in-human biological movements to be suspended in its flow (Keil and Ali 2007).

Vectors are used not just before the 'building event' (Jacobs 2006; Massumi 2002) of terminal construction, or *in* the event of air-traffic control communications. For instance, flight-track monitoring systems are now being used to monitor the behaviour of pilots and aircraft in order to scrutinise them later. Should an annoyed local resident complain about the volume levels of a recent flight, or the proximity of an aircraft to their residence, the airport's environmental manager may trace the mobility of that aircraft back to its path and therefore its point in space/time the resident is referring to. Using that vector, the environmental officer may even trace it back to the nearest monitoring position which he may call forth from his computer desk in the airport building. Digitised sound recordings taken by monitoring devices have their own modems which allows them to transmit their information and a recording of the offending noise to the airport should the

Environmental Manager wish it. The logic of the vector, in short, is also the logic of position.

Link

Vectors constitute and exceed representations. It should be said that such an idea is not necessarily new even within a field such as Geography, it is just perhaps overlooked. For George Zipf, one of the fathers of the field of social physics in the 1950s who greatly influenced both geographers, economists and sociologists at the time, an individual – say an individual called John – does much more than draw lines, or what Zipf called 'paths'. For Zipf (1949, 3), John 'is a *set of paths*' while he is also 'a unit' that '*takes paths*'. Zipf (1949, 3) goes on, 'If we restrict our attention to John as a set of paths in reference to which matter-energy moves into John's system, through John's system, and out of John's system, we note that there is nothing in this transient matter energy that can be called permanently 'John'. The realisation that John is nothing more than a set of paths is useful. As is the concept that John has no permanency. He has no whole position for he is located or distributed along the network.

Another and infinitely more famous John than Zipf's is of relevance here: the Beatle, John Lennon. In 2001 Liverpool Airport became Liverpool John Lennon Airport in honour of the city's most famous son. This event became what Mackenzie Wark (1994) might describe as a media *event*. As well as playing host and distributor to numerous vectors of aircraft, people and things, the airport became the content of media vectors. The airport traveled along the conduits of the internet, television and satellite communication systems, newspaper and magazine publications. The vectors connected the airport to places across the globe that its flights may or may not have flown to, and rather than people, 'images, and sounds, words, and furies' were shuttled between them (Wark 1994, 11–12). Rolling Stone magazine, numerous Beatles weblogs, listservs and websites picked the story up along the passage of its distribution.

In this way the airport and lots of other things occupy extremely extensive spatialities (Mitchell 2004), their vectors criss-cross the globe to saturation point. Like the biological transmissions described above, these vectors are carriers for one another. Liverpool Airport as a brand name piggy-backed the brand of John Lennon and his wife, Yoko Ono, whose vectors are considerably more pervasive and extensive than the fastest jet.

In another context, Liverpool Airport resides in the fiction of the horror writer Clive Barker. Viewing the 1956 Liverpool Airshow from a cornfield nearby, Barker observed the famous Birdman Leo Valentin (so named because of the wings sewn into his sky-diving suit) tumble from the aircraft. Caught up in his parachute cables Valentin 'roman candled' and fell to his death. 'There are no other events in my early life which carry quite the primal force of Leo Valentin's fall' (1999, 10) Barker writes. Valentin's fall often appears in the spaces of Barker's novels, embodied in the familiar character of a winged man or a birdman. In this sense the vector of Valentin's downwards fall at Liverpool has persisted. Its 'power to insist itself upon' the imagination seared the fall onto the 'rock' of Barker's skull as he puts it – rendering a mould or a blueprint 'from which all manner of other tales and pictures would in time be derived' (1999, 10). The line of the Birdman repeats itself in Barker's fiction. The directional vector of the fall pervades his stories allowing Valentin's movement to live on in narrative.

The vectoring of airports poses some issues. It is not simply the problem that airports change, in Henri Bergson's (1911) sense of evolution. This routinely happens through practices of maintenance, repair, refurbishment and expansion (Thrift and Graham 2007). Rather, it refers to the difficulty in locating airports. They do have relatively fixed coordinates, but the vectoring of airports ensures it is hard to know where the airport ends and something else begins (Merriman 2007, 109–10).

Airport sounds penetrate the airport boundaries. We have seen how nearby residents send forth their own vectoral communications in their complaints to the airport. The airport itself is continuously represented by envoys and ambassadors (Serres 1995) that populate the city and surrounding area. Adverts and aircraft distribute the airport beyond itself. The bus service known as the Liverpool 'flight link' attaches more vectors of communication and presence between the airport and the city, or lines of transit to Manchester city centre by moving along the M62 motorway in-between. Wandering around the nearby National Trust property, Speke Hall, one comes across ruins and relics of where the airport used to be (Edensor 2005). The old Liverpool Airport at Speke still exists – now as a hotel. One finds remnants of the airport's taxi ways and pavements that linked it to the runway now used by the new terminal building. The airport's infrastructure, once a pathway for the facilitation of aerial mobilities now provides a path-way with an end – a route for owners walking their dogs and couples strolling alongside the Mersey.

Just as Nick Barley (2000) is troubled by the difficulty in finding London, airports such as Liverpool's are at once easily definable by simple coordinates and even a name, yet also un-locatable – stretched-out by transportation and communication vectors.

Point-Past-Future

As I mentioned at the beginning of this chapter, many vectors are difficult to perceive as a whole. If one figuratively stands alongside a vector, it may cross the presentness of their experience as they live it, just as the aerial contrails pass out of the dimensions of the image at the top of this chapter. The spatial and temporal reach of a vector can extend beyond the now to where it was, where it has gone to and, probably, where it will go. In this section I suggest that this past, present, and futurity exists as more than an abstraction, or a material trace, but they constitute embodied dimensions of common airport experiences.

Let us consider Peter Fleming's (1982) all-too-real description of what it is like waiting at an airport terminal for one's plane to board. Fleming tries to probe the fears and anxieties of waiting. The seemingly endless periods may be even felt as boredom and introspective behaviour. Those of us familiar with contemporary air-travel are probably well-versed with the time-stilled and space-slowed stretches endured – to use Ben Anderson's (2004) terms. Normally characterised by incredible social inactivity, passivity and inattention (see also Hirschauer 2005), airport boredom may be broken by moments of activity and anxiety. Fleming puts forward the question: 'which is worst'? In the first instance, is it the waiting for one's plane to arrive? These experiences are now-familiar given the 25 minute turnarounds endemic to low-cost air travel and airlines such as Ryanair, easyJet and South West. What sorts of anxieties are afforded in those moments waiting at the gate scouring the horizon for the aircraft to come? Or on the other hand, is it the case when the aircraft is there, waiting on the tarmac for eventual departure, yet solemnly immobile and delayed? Fleming writes:

> If the aircraft is actually present – immobile indeed, but glossy and reassuring in appearance – it is generally grounded by an unfavourable weather report, by a mechanical breakdown or – in wartime or in some Oriental countries – by the non-arrival of an important passenger. (1982, 205)

What Heidegger (1977) describes as the aircraft's 'ready-to-handness' is constrained and delimited as potential lines of escape are cut short by various kinds of organisational or technological breakdown; by a fuel pump failure, by a maintenance check, by malfunctioning toilets. For Fleming, this kind of practice is incredibly testing for the character, involving what he describes as a 'peculiar blend' 'of boredom with anxiety, of false hopes with unnecessary fears' (1982, 206).

The contingent complexity of air-travel makes these internalised and virtualised vectors that much more urgent. As Fleming asserts, the uncertainties associated with flight delays are continuously shifting, 'these causes of delay are bound to be transient'. The waiting passenger may become sensitive and attuned to these contingencies as they study the movements of weather patterns and cloud formations. Past weather variations suddenly take one's interest, 'But gradually, as we wait, the omens become less and less encouraging. The sky clouds over, and some vile fellow says that if you get bad weather in this part of the world it generally lasts for a week' (Fleming 1982, 206). Others may observe the efficiency of baggage handlers loading on the tarmac, and the caution of staff checking the plane and loading on fuel.

The anticipatory and hopeful attendance to vectors form alongside the formation of an immanent passenger-body. We might think of these lines as potentials than run along the vector, what we can refer to once again as 'desire lines' (Tiessen 2007); they have direction and act in the way Lingis (1998) would describe as directives. People prepare themselves to go as passports and boarding tickets are placed in hand; as handbags are adjusted accordingly. Upon any hint or sign of departure the immanent action is released, 'The cowling on the port engine is replaced and we gather our belongs hopefully together, one inexperienced traveller even knocking out his pipe' (Fleming 1982, 206). Bags are shut, newspapers are folded, people change their postures, others stand up. Preparatory states of readiness (Crandall 2006) appear to unlock. The action once held at bay is now let go. Mimicry sees these actions rippling down the attentive mass of passengers like a shockwave. They are now go-ing. But if we return to Fleming's scene these movements are quickly stalled, 'they rev her up, although to us the roar sounds perfectly satisfactory, the crew shake their heads and the cowling comes off again' (1982, 206). Thwarted again by another malfunction, the passenger's attitude changes. The aircraft's imminent departure, which was mirrored in the passenger's shifts, changes its meaning and the attitude of the passengers moves once more, 'The great silver machine, whose air of being poised for flight once seemed encouraging, becomes a hateful spectacle, a symbol of futility and disillusionment' (Fleming 1982, 205–6).

These anticipatory or virtual vectors are distinctly embodied. Following David Bissell (2007), we know how even the most immobile and fixed of states can belie and hide a storm of thoughts, emotions, sensations and feelings. But they are not held within the interior of the body, nor do they solely come from there. These sorts of states appear to run along the line. Anticipatory states invoice a present embodied in bodily postures, attitudes and movements. Fleming goes on to describe,

> At the distant sound of engines our spirits rise, we feel that we are practically airborne already, but the plane, if it lands, turns out to belong to some other service, and we resume our moody vigil with the worst possible grace. The stubble on our chins, the breakfast which was bolted hastily in the dawn and makes us feel dyspeptic, remind us of how early we were dragged from that state of oblivion which we now crave on mental as well as physical grounds. We long to deface the posters in the waiting-room which extol the celerity, the comfort, the almost Elysian delights of travel by air. We long, above all, to do what we came here to do: to fly. (1982, 207)

A vector blocked sends these affects into a feedback loop. They cannot be released because the body cannot go any further. Irritation and aggravation may thrust the passenger's feelings outwardly towards other passengers and airline staff. Withdrawal is just as likely too. Despondency and sadness may see bodies acquiesce into states of passivity.

Nerve-Centres (of Calculation)

Airports obviously transmit information. Vectors of communications centre and pass through them which they have to coordinate and control (Fuller and Harley 2004). It is for this reason that airports have to position themselves in such a way that they can capture, receive and send out these flows. Such a relation is cast in the efforts of the first airports to orientate bodies to them. The design of Liverpool Airport's first control tower speaks to these issues as the control tower was designed much like a lighthouse (Butler 2004). Without the terminal buildings which were completed some three years later, the building jutted up from the ground to some 70 feet. And like a lighthouse, the control tower sent out vectors of information such as a beacon light. Yet it was also built to receive flows of information.

The building and its various technologies brought the vectors it needed to receive and send out close-to-hand and visible. It is in this way that the building might be understood through what Latour (1999) describes as a 'centre of calculation'. The control tower made the movements of aircraft knowable by transforming their paths and lines into diagrammatic representations. Lines of literal flight were drawn in order to plot the routes and position of aircraft. In a reversal, incoming aircraft would then align themselves to the radio beacon which took the physical and representational shape of a vector – a line of transmission running along the airfield's path. These early vectors enabled pilots and control officers to fix and grab hold of their positions.

In drawing and representing vectors the complexity of the airport, which was considered a system or a nervous system, became legible, knowable and manageable. The vectors brought mobility near-to-hand so that it could be managed. As Latour writes of the astronomers' chart: 'each of them brings celestial bodies billions of tons heavy and hundreds of thousands of miles away to the size of a point on a piece of paper' (Latour 1987, 227), a similar apparatus is supplied within and through the control tower. Charts and diagrams made mobility knowable; flows of movement were translated into fixed and static information to be stored on graphs and charts.

The bodies of control tower officers were interfaced with these vectors. With a quick turn of the control officer's swivel seat he could face the compressed air cylinder system behind him to pass and receive messages through the rest of the tower, all without moving from the spot. The tower's ultimate aim was to bring or suck in all these flows as close as possible to the control officer's body and other members of the tower's staff. According to a report at the time, 'The control officer on duty in the tower will be able to control the whole mechanism of the airport'.[2] The officer would command the comings and goings of aircraft. Much was made of the immobility and technological prowess of the tower, the centre point from which all information flowed and was concentrated. A report stated, 'Seventy feet

---

2   Ibid.; 'Speed up in airport plans', *Liverpool Evening Express,* 19th February 1937. Liverpool Records Office (LRO thereafter).

above ground level in the octagonal control room, the control officer, without moving from his seat, has an unbroken view in all directions, and the intricate apparatus can be operated at the touch of a button'.[3] Reports boasted of the, '... most up-to-date control yet devised, enabling the control officer to keep an eye and an ear on all that occurred within site and range of the aerodrome'.[4]

With its commanding views of the whole of the airport, the new tower was seen as the brain or the 'nerve centre' of the airport machine. The widespread introduction of electricity into the public sphere encouraged a new 'linguistic currency' of wires, cables, flows, connections and energies (Armstrong cited in Thrift 1996: 272) into descriptions of the 'electric' airport. Here, the body became transfigured into the circuitry of the terminal. Bodily agency was distributed to all parts of the airfield and beyond as the airport itself was increasingly conflated with the human body, allowing their commensurability. Press articles and popular commentaries consistently used bodily metaphors to describe its function. As one article stated, 'The Control Tower may be compared with the spinal column and nerve centre of the human system', through the 'amazingly intricate electrical apparatus' that allowed the building to work.[5]

Conclusion: For Vectors

If we recall the brief discussion concerning time-geography at the beginning of this chapter, I mentioned how Torsten Hägerstrand was well aware of the limitations of his approach – how one might want to get beyond the line or look through the paths people drew. For Hägerstrand, the vector was the tip of the iceberg-sized volumes of experience submerged in his diagrams. We are reminded of this argument by Tim Cresswell, who makes a similar point in his *On the Move* (2006). In this instance, Cresswell draws a simple illustration drawn below:

A--------------------->B

For Cresswell, the diagram is representative of movement and not mobility. It is an abstracted movement characteristic of the way mobility has been treated as a remarkably understudied component of the social world. Cresswell wants to 'unpack' this line and not take it for granted. But while Cresswell looks at lines as an abstracted component of a fuller reality – a movement inscribed with meaning – it does seem to me that the line itself needs looking at more closely.

In fact, I would suggest that many approaches towards mobility are a lot closer to Cresswell's diagram than is at first obvious. Efforts to unpack and get

---

3   'Opening of tower and hangar', *Liverpool Daily Post,* 12th June 1937. LRO.

4   Ibid.

5   'The Liverpool Airport and its Nerve Centre: The Control Tower', *The Liverpolitan,* July 1937. LRO.

underneath this diagram – if you will – have a tendency to break the line up. Just like the dashed-line that makes up his diagram (a representational error of the publishers) we are left with a bop, bop, bop, bop of moments and not vectors. The lines of mobility are broken and comprehended at various instants along the path of their wider displacement. Examining a mobile body moving through an airport for instance, is to examine a mobility within an enormous wider context of mobilities travelling to the airport and aeromobilities across the globe. I liken this sort of thinking to what Marcus Doel (1999) describes as 'pointillism'. Instead of taking a '-' or a '.' can we understand these moments and points more extensively? Can we take the two-dimensional image and extrude it outwards through the page?

One benefit of vectors is therefore scalar. Vectors dislocate and as Alliez states (see Thrift 2006), they *pluralise* the world. In the case of the airport they reveal an extensiveness that crosses and goes beyond lived experiences. Establishing where exactly Liverpool Airport is through vectors becomes incredibly problematic. Being careful not to reproduce too many of the myths of globalisation and homogenisation, the airport crops up in lots of different places: along motorways; through an estate owned by the National Trust; in television broadcasts; and in the pages of horror fiction. An examination of vectors reveals direction, origin, constraints and limits, an airport squeezed and shuttled along particular pathways or corridors of travel (Lassen 2006). Understanding airports according to their topological position as a node connecting a network, thus, disguises their existence along the lines of the network itself.

Examining vectors highlights something so simple about mobility it is often overlooked. This simple fact is direction. We have seen how the very different directional headings of falling and take-off have very different registers in these examples. While mobility has no pre-existent meaning, different vectors are made sense of in very different ways.

We have seen how vectors intersect the representational and non-representational. Vectors are both ways of representing mobility in order to make it sensible and controllable. In another way, vectors are means to orient oneself to mobility. But vectors are also felt and constitutive parts of mobility. The physical metaphors of pipes and infrastructures evoked in recent writings concerning 'affect' (McCormack 2006; Thrift 2004; see Thien 2005 for a critique) resemble the sorts of vectoral threads and pullies described in this chapter, with definite directions of source and destination. These threads *move* people in the sense that we may understand emotions and feelings as not bound to an interiority, but distributed extensively and brought forth into action. Feelings may even alter and interrupt the direction of their own paths as the vectoral line may turn in on itself to create a loop.

'Where do those tubes go?' asks Lincoln Six Echo. Let us find out.

References

Adey, P. (2004), 'Secured and Sorted Mobilities: examples from the airport', *Surveillance and Society* 1, 500–19.

Adey, P. (2006), 'Airports and Air-mindedness: spacing, timing and using Liverpool Airport 1929–1939', *Social and Cultural Geography* 7, 343–63.

Anderson, B. (2004), 'Time-stilled space-slowed: how boredom matters', *Geoforum* 35, 739–54.

Armstrong, T. (1992), 'The Electrification of the Body', *Textual Practice* 8, 16–32.

Barker, C. (1999), *The Essential Clive Barker: Selected Fictions* (London: Harper Collins).

Barley, N. (2000), 'People', in Barley, N. (ed.) *Breathing Cities* (Basel: Birkhauser).

Bergson, H. (1911), *Creative Evolution* (New York: H. Holt and Company).

Bissell, D. (2007), 'Animating Suspension: Waiting for Mobilities', *Mobilities* 2, 277–98.

Butler, P.H. (2004), *Liverpool Airport: an Illustrated History* (London: Tempus Publishing).

Crandall, J. (2006), 'Precision + Guided + Seeing', *CTheory*, http://www.ctheory. net.

Crang, M. (2002), 'Between places: producing hubs, flows, and networks', *Environment and Planning A* 34, 569–74.

Cresswell, T. (2006), *On the Move: the Politics of Mobility in the Modern West* (London: Routledge).

Cwerner, S.B., Kesselring S. and Urry J. (2008), *Aeromobilities* (London: Routledge).

Deleuze, G. and Guattari F. (1988), *A Thousand Plateaus: Capitalism and Schizophrenia* (London: Athlone Press).

De Neufville, R. and Odoni A.R. (2003), *Airport Systems: Planning, Design, and Management* (New York: McGraw-Hill).

Doel, M.A. (1999), *Poststructuralist Geographies: the Diabolical Art of Spatial Science* (Edinburgh: Edinburgh University Press).

Edensor, T. (2005), *Industrial Ruins: Space, Aesthetics and Materiality* (London: Berg).

Fleming, P. (1982), 'With the Guards to Mexico' in Kennedy, L (ed.) *A Book of Air Journeys* (London: Harper Collins), 202–08.

Fuller, G. and Harley, R. (2004), *Aviopolis: a Book about Airports* (London: Black Dog).

Graham, A. (2003), *Managing Airports: an International Perspective* (Oxford: Butterworth-Heinemann).

Graham, S. and Thrift N. (2007), 'Out of order – Understanding repair and maintenance', *Theory, Culture & Society* 24, 1–18.

Gregory, D. (1985), 'Suspended animation: the stasis of diffusion theory', in Gregory, D. and Urry, J. (eds) *Social Relations and Spatial Structures* (Basingstoke: Macmillan), 296–336.

Haggett, P. (1965), *Locational Analysis in Human Geography* (London: Edward Arnold).

Heidegger, M. (1977), *The Question Concerning Technology, and other Essays* (New York: Harper and Row).

Hirschauer, S. (2005), 'On doing being a stranger: The practical constitution of civil inattention', *Journal for the Theory of Social Behaviour* 35, 41–67

Jacobs, J.M. (2006), 'A geography of big things', *Cultural Geographies* 13, 1–27.

Keil, R. and Ali, H. (2007), 'Governing the Sick City: Urban Governance in the Age of Emerging Infectious Disease', *Antipode* 39, 846–73.

Lassen, C. (2006), 'Aeromobility and work', *Environment and Planning A* 38, 301–12.

Latour, B. (1987), *Science in Action: How to Follow Scientists and Engineers Through Society* (Milton Keynes: Open University Press).

Latour, B. (1999), *Pandora's Hope: Essays on the Reality of Science Studies* (Cambridge, MA: Harvard University Press).

Lingis, A. (1998), *The Imperative* (Bloomington, IN: Indiana University Press)

Lowe, J.C. and Moryadas S. (1976), *The Geography of Movement* (Boston, MA: Houghton Mifflin).

Massumi, B. (2002), *Parables for the Virtual: Movement, Affect, Sensation* (Durham, NC: Duke University Press).

McCormack, D. (2006), 'For the love of pipes and cables: a response to Deborah Thien', *Area* 38, 330–32.

Merriman, P. (2007), *Driving Spaces* (Oxford: Wiley-Blackwell).

Mitchell, W.J. (2004), *M++ The Cyborg Self and the Networked City* (London: MIT Press).

Pascoe, D. (2001), *Airspaces* (London: Reaktion).

Serres, M. (1995), *Angels, a Modern Myth* (Paris: Flammarion).

Solnit, R. (2003), *River of Shadows: Eadweard Muybridge and the Technological Wild West* (New York: Viking).

Thien, D. (2005), 'After or beyond feeling? A consideration of affect and emotion in geography', *Area* 37, 450–4.

Thrift, N. (1996), 'Inhuman geographies: landscapes of speed, light and power', in Thrift, N. (ed.), *Spatial Formations* (London: Sage).

Thrift, N. (2004), 'Intensities of feeling: Towards a spatial politics of affect', *Geografiska Annaler Series B* 86, 57–78.

Thrift, N. (2006), 'Space', *Theory, Culture and Society* 23, 139–46.

Tiessen, M.P. (2007), 'Urban Meanderthals and the City of "Desire Lines"', *CTheory* http://www.ctheory.net.

Urry, J. (2000), *Sociology Beyond Societies: Mobilities for the Twenty-First Century* (London: Routledge).

Wark, M. (1994), *Virtual Geography: Living with Global Media Events* (Indianapolis: Indiana University Press).

Wells, A.T. and Young, S. (2004), *Airport Planning and Management* (New York: McGraw-Hill).

Zipf, G.K. (1949), *Human Behavior and the Principle of Least Effort; an Introduction to Human Ecology* (Cambridge, MA: Addison-Wesley Press).

Chapter 10

Immigration Stations:
The Regulation and Commemoration of
Mobility at Angel Island, San Francisco and
Ellis Island, New York

Gareth Hoskins and Jo Frances Maddern

Introduction

In the writings of History, Geography, and Social Theory the United States has long been discursively tethered to concepts of mobility. Zelinsky epically remarked how 'The love of change and of all forms of mobility, an innate restlessness, is one of the prime determinants of the structure of the American national character' (1973, 53). Movement, travel, and immigration are key tropes in the national narrative economy (Kouwenhoven 1961; Boorstin 1966; Baudrillard 1988); they are crucial mechanisms through which individuals overcome nature as wilderness (Turner 1947) and cultivate community out of diversity (Agnew and Smith 2003). Despite the United States' representational affinity with the concept of mobility, making movement meaningful necessarily depends on a host of material social practices emerging from and embedded in grounded spaces (Cresswell 2001; Crang 2002; Blunt 2007). In the following pages we explore two of these grounded spaces where one kind of movement, immigration, is invested with meaning through the practices of regulation and commemoration.

In the late nineteenth and early twentieth centuries Ellis Island in New York and Angel Island in San Francisco were two among seven inspection centers strategically located around the United States in a federal bureaucratic complex designed to regulate the movement of immigrants. We contrast the construction of mobility as threat in this period with the construction of mobility as full of promise in the spaces as they exist today – as national heritage sites of secular pilgrimage generating thousands of visitors annually. Our discussion synthesizes two separate research projects. The first explored themes of mobility and memory and was conducted at Angel Island San Francisco. It entailed an eleven-month ethnography, extensive documentary research, and a series of interviews with those closely associated with the site's operation (Hoskins 2005). The second documented the restoration of the museum of immigration at Ellis Island. It involved forty interviews with producers over a period of eight months and an analysis of National Park Service archives in New York, Washington, Boston and West Virginia (Maddern 2005).

Angel Island is regularly portrayed as a space of immobility and imprisonment (Takaki 1989; Kingston 1980; Limerick 1992; Daniels 1997) whereas Ellis Island is popularly understood as a set of 'golden doors', an almost mythological site where multicultural America was formed – as represented in films including *The Godfather* (1972), *Hitched* (2001), novels *Liberty Falling* (2000), and music including *Ellis Island* by the Irish Tenors (2001). Although similar in many respects, these spaces are internally complex and ambiguous, drawn together by a series of laws that find unique expression according to their particular geographical characteristics. Bringing these two examples together allows us to highlight the nuances missed in oppositional representations of Angel Island and Ellis Island and how this opposition has marked parallels with theoretical readings of the spaces of mobility and immobility more generally.

Theorizing Mobility

Islands have a tenacious hold on the human imagination due in part to their provision of stability amongst the watery chaos of their surroundings (Tuan 1974, 118). Islands have a peculiar relationship with the concept of mobility. Gillis (2003, 2004) notes how islands can represent both separation and continuity, isolation and connection: 'The idea of the island brings with it at once the notion of solitude and of a founding population ... islands inhabited by human beings are never enclosures only: they are crossroads, markets for exchange, and while sail remained the mode of transport they were essential and frequent stopping off points for re-provisioning' (2004, 33). US Immigration Stations Ellis Island and Angel Island are apposite examples of mobility's ambivalence since they seem to mirror enduring oppositional binaries associated with movement: movement 'as something to fear' and movement 'as something to celebrate'. Ellis Island is portrayed as a place that is 'likely to connect with more of the American population than any other spot in the country' and as 'one of the most popular tourist destinations in the country' (The Statue of Liberty-Ellis Island Foundation Inc. website) since apparently half of all Americans are able to trace an ancestor who once passed through its gates. Of course, such universalist statements omit the particular experiences of those excluded in one way or another: African-Americans who arrived as slaves, Hispanic migrants from the US-Mexican border, and Asian migrants coming from the West.

Angel Island has long been declared the 'Ellis Island of the West' (Bamford 1917), but at the same time, its absolute differences are highlighted. As Wallenberg's entry in Stolarik's essential inventory of 'other ports of entry into the United States', *Forgotten Doors* makes clear: 'Angel Island was therefore not the Ellis Island of the West, a receiving point for most west coast immigrants of all nationalities. Instead it was a peculiar product of the Chinese Exclusion

Act whose primary purpose was to control and restrict immigration' (Wallenberg 1988, 149). The direct similarity or direct distinction between these two spaces disambiguates a continuum as if the condition of mobility has to be either wholly positive or wholly negative. This opposition of extremes abounds in all manner of historical accounts when applied to the immigrant whose presence is said to provoke contradictory responses of *xenophobia* and *xenophilia* (Honig 2001). On the one hand the immigrant is conflated with the notion of threat, harbinger of disease, danger and instability – an expression of deep rooted anti-immigrant sentiment that precipitated the National Immigration Act (1921), National Origins Quotas (1924) and the Chinese Exclusion Acts that ran through the late 1800s and early 1900s up to their rescinding in 1943. On the other hand, the character of the immigrant plays a more positive role reconfirming belonging in a multicultural cosmopolitan nation. Honig calls this particular caricature the 'supercitizen' who is crafted by media reports and citizenship ceremonies to highlight the nation's desirability:

> ... the liberal consenting immigrant addresses the need of a disaffected citizenry to experience its regime as choice-worthy, to see it through the eyes of still enchanted new comers ... the immigrant's decision to come here is seen as living proof of the would be universality of America's liberal democratic principles. (Honig 2001, 75)

It is possible to draw parallels between the ambivalence of the immigrant as a mobile subject and the ambivalence of attempts to understand mobility as a working intellectual concept. In social theory tensions remain between sedentarist and nomadic perspectives. Sheller and Urry (2006) argue that social science has largely been static and 'a-mobile' where stability is taken as the norm, as fundamental, and mobility a typically dysfunctional force threatening the authenticity of place and rootedness. We can see this sentiment played out in a plethora of geographical writings (Kunstler 1994; Augé 1995). Conversely, mobility has been celebrated in postmodern writings (Clifford 1997) that valorize movement to an extent that the difference between kinds of movement is underplayed, depoliticized and flattened out (Crang 2002). Recent writings on the concept of mobility and contributions in this volume, have challenged us to move beyond 'movement versus stasis' antagonisms by focusing on the actual content of movement as it is experienced and undertaken. As Cresswell argues: 'The way forward is signposted by theories of practice which serve to destabilize notions of place and space' (2003, 20). Attention to practice allows us to consider the thresholds, gateways and portholes (often maligned as non-places) through which people move as the very spaces where movement is codified, regulated, produced and experienced as mobility. Theories of practice (Bourdieu 1977; de Certeau 1984) help us to focus on the role of space in embodied encounters between different individuals and the norms and knowledges that shape those encounters; they help us to see how mobility is not only abstract and metaphorical but central to the emergence of particular kinds of corporealities that become associated with particular material spaces (Martin 1997; Tyner 2006).

A focus on the spatial groundings and physical infrastructures through which mobilities are produced allow us to counter those mobile epistemologies directed toward displacement, intersection and interaction which often elide those very processes that make movement meaningful. While we acknowledge that the production of mobilities at Angel Island and Ellis Island are not confined to their respective coastlines – that through practices of regulation and commemoration the mobilities configured are certainly networked and geographically extensive – these spaces nevertheless offer coherent and relatively fixed contexts to examine how the production of mobility takes place. With a principal focus on spaces, the objective is not to tether the concept of mobility to particular discrete locales but rather to suggest how both locales and the practices that occur within them shape the way mobility comes to make sense.

We continue our discussion with three sections. The first shows how regulating the movement of international migrants provoked the creation of a new kind of space, the immigration station, with new kinds of characteristics. The second section jumps to the present day and examines how a commitment to remember the immigrant experience as a shared national trait has meant the wholesale conversion of these island locations in ways that create new kinds of mobilities that commemorate those they supersede. The final section explores the role these islands play in the configuration of future mobilities where selective renderings of the past are used instrumentally to shape meanings given to movements yet to come.

Regulating Mobility

From their inception, the US Immigration Stations at Ellis Island and Angel Island played a pivotal role in defining mobility as an external and foreign threat to 'the good order of certain localities' (Chinese Exclusion Act, 1882, Preamble). Here, inspectors, eugenic researchers, commissioners, union leaders, elected officials, lawyers, brokers, ambassadors, and United States Public Health Service employees became embroiled in an institutional web of nativist ideology and racist discrimination where mobility became *the* condition through which the nation's foreign born and domestic populations were administratively controlled. Martin (1997) employs Bourdieu's notion of *habitus* to show how bodies and spaces are 'ethnicised' to invoke acceptable and unacceptable subjects. She argues, 'the space of the nation is thus characterised as delineated by a bodily way of being in the world – the body as lived actually constitutes the nation. The nation is not only or even primarily 'imagined', abstract or metaphoric, but is part of materiality, corporeality, political economy of bodies' (Martin 1997, 108). The US immigration stations of Angel Island and Ellis Island were charged with inscribing value on the bodies passing through both quantitatively, in terms of monitoring national admittance quotas, and qualitatively, in terms of individually assessing immigrants' on-site movements for signs of eligibility.

On the West coast, US Immigration Station Angel Island operated between 1910 and 1940 to process approximately one million immigrants including Japanese, Koreans, Filipinos, Asian Indians, Mexicans, Spanish, Portuguese, Russians, and South and Central Americans. The majority of those detained for questioning, however, where of Chinese decent. These Chinese migrants were subject to a series of laws (Chinese Exclusion Act 1882, 1902; Geary Act 1892; The Gentlemen's Agreement 1907; Immigration Act 1917; National Origins Act 1924) that pathologized Asians generally, but particularly Chinese individuals, on the basis of their 'race' and social class.

On the East coast, US Immigration Station Ellis Island operated between 1892 and 1954 to process more than twelve million immigrants mostly of European descent. While procedures at Angel Island were geared towards assessing relatively fixed physiological features using photographic identification and visual readings of immigrants' bodies (Pelger-Gordon 2006), at Ellis Island scrutiny was directed more towards how these features became dynamic; how movement, expression, gait, poise, and posture might suggest deficiencies in sanity, intelligence, literacy, and health. Importantly, the spatial orientation of the processing environments at Ellis Island and Angel Island evolved to support the inspection practices taking place. An immigrant's right to mobility on a global scale, therefore, became a condition of the micro-mobilities he or she enacted on the ground under the gaze of immigration staff.

The impetus for the construction of dedicated immigration facilities at Angel Island and Ellis Island came alongside a tightening of immigration law when existing mainland facilities were deemed inadequate. Offshore locations could enhance security, prevent collusion between immigrants and US citizens, and segregate immigrants with communicable diseases. In a letter to the Commissioner General of Immigration, Washington D.C, Commissioner North argued:

> I understand the Chinese community aided and abetted by their white lawyers have taken every possible step to thwart the opening of the station at Angel Island. To properly comprehend this one must know that almost the entire Chinese community of San Francisco and surroundings are engaged, whenever the opportunity offers, in the fraudulent importation of their fellow countrymen. When such importations have been held, at the mail dock or detention shed, it has been absolutely impossible to prevent communication between such arrivals and their friends on shore, thereby enabling them to perfect their stories and making it doubly hard for this service to prove fraud. (North 1910, 10)

The construction of facilities on both Ellis Island and Angel Island took explicit instruction from the Bureau of Immigration. Architectural design (Italian Renaissance Revival style at Angel Island and Beaux-Arts at Ellis Island) functioned to emphasize the imposing authority of the state and to instil a sense of professionalism and scientific objectivity amongst employees. Walter Matthews, involved in the design of facilities on Angel Island, described the character of the site as a 'parade ground' (Davidson and Meier 2002). Of Ellis Island, Kraut remarked that: 'Immigration Bureau inspectors and US Marine Hospital Service physicians saw Ellis Island's ornate turrets as towers of vigilance from which they dutifully guarded their country against disease and debility' (Kraut 1994, 52). From these new spaces of control, the usually abstract world of the geo-political was distilled down to proximal embodied encounters where immigration staff could reinforce prejudices of the Italians as poor, the Poles as weak, the Chinese as opium addicts or the Greeks as unhealthy. Ngai (1999, 69) notes that The National Origins Act of 1924 (43 stat. 153) was based on 'a constellation of reconstructed racial categories in which race and nationality – concepts that had been loosely conflated since the nineteenth century – disaggregated and realigned in new and uneven ways.' The form in which this realignment took place cannot be explained without direct reference to the on-the-ground practices of inspection occurring within specific processing environments. Discursively constituted categories of race, class, gender, disease, and sanity, easily defined on paper in the instruction manuals and diagnostic handbooks referred to by immigration staff, became confounding, indistinct, and bewildering when having to be implemented on the ground.

In their quantitative analysis of immigration processing and detention figures, historians Barde and Bobonis give some hint to the ambiguity involved in transferring discourse to practice, 'it is possible that there existed a large amount of idiosyncratic day-to-day variation in the decisions' (2006, 120), and observed that much of the variation in the administering of Immigration policy 'seems random; that is, uncorrelated with changes in official national policy. This is probably due to the proclivities of the staff at Angel Island and changes in local political pressures to enforce exclusion or not' (2006, 131). Mass immigration through Ellis Island also posed an administrative challenge. This is evident in a 1903 Public Health Service Book of Instruction which sought to map out a geography of the insane body to aid officers in making decisions on admission:

> Insanity is a deranged and abnormal condition of the mental faculties ... In the
> case of immigrants, particularly the ignorant representatives of emotional races,
> due allowance should be made for demonstrations of excitement, fear of grief.
> (quoted in Kraut 1994, 67)

In addition to Public Health Service guidelines each doctor or inspector had their own way of identifying the inadmissible body. Often such diagnoses were little more than hunches as this advice from a doctor in 1925 demonstrates:

> There is a no more difficult task to detect poorly built, defective or broken
> down human beings than to recognize a cheap or defective automobile ... The
> wise man who really wants to find out all he can about an automobile or an

immigrant, will want to see both in action, performing as well as at rest, and to watch at a distance as well as to scrutinize them close at hand... A man's posture, movement of his head or the appearance of his ears ... may disclose more than could be detected by puttering around a man's chest with a stethoscope for a week. (Stafford, 1925 in Kraut 1994, 67)

Thus the responsibility for exclusion shifted from detached policy makers to on-site physicians and inspectors. Medical knowledge as it developed through the testing and observation of migrants was not value-free but linked to a politics of mobility designed to maximize the economic productivity and minimize public welfare costs.

At Angel Island the ambiguities of assessment opened the way for fraud. Chinese immigrants entered illegally using false documents and various disguises. Immigration guards received bribes to rush through paperwork or provide answers to identity tests. One instance of this subterfuge is highlighted in a 1917 investigation by J.B Densmore which ended with the dismissal of 25 employees from the immigration service including seven out of the eleven inspectors hearing Chinese cases. The investigation reported an array of clandestine services involving the substitution of photographs in immigrant files charged at a rate of US$200, and the theft and copying of interview transcripts for a fee of between US$100 and US$150 (Lee 2003, 200).

Policing the Chinese Exclusion Act, a bill that combined racial and class dimensions as criteria for exclusion created unresolvable enforcement dilemmas requiring an array of ad hoc procedures invented by creative immigration staff (Calavita 2000). Clues towards eligibility could be assessed through physical markers and biological traits. Calluses on hands would indicate that an immigrant presenting himself as a merchant, say, might in fact be an unskilled labourer excluded under the law. Anatomizing the law through inspection practices reached its apogee in 1903 with the temporary adoption of the Bertillion System (a data-set of relative measurements of subjects' fingers, feet, forearms, teeth and skull width) that originated in France to assist in the positive identification of criminals. However, bodily indicators could only be taken so far, particularly when the ambiguities of the flesh failed to correspond directly to a range of 'objective' categories and markers. As a necessary supplement, interrogation sessions were established to verify immigrants' identity against groups exempt from the exclusion laws: students, teachers, ministers, merchants, tourists and their dependents.

The conflation of mobility with threat was illustrated through an immigrant's physical confinement and their carefully choreographed movement through the facilities. After disembarking from the wharf at Angel Island possessions would be removed, and immigrants striped naked and inspected by hospital staff (Markel and Stern 1999; Shah 2001). A period of detention would follow before a hearing was held. During this time men were separated from women and children, and nationalities (usually Chinese and Japanese), divided. To mitigate against possible collusion, no contact with mainland friends or relatives was allowed. A letter by former assistant surgeon M.W. Glover of the Public Health and Marine Hospital Service to the Acting Commissioner of Immigration in San Francisco regarding sanitary conditions of the station at Angel Island highlights the restrictions placed upon daily movement.

> The aliens are locked in the dormitories except when allowed out in the small enclosure known as the 'recreation ground'. In unfavourable weather this means that several hundred aliens are kept in these rooms continuously ... The windows are barred and locked. There is only one exit for all four dormitories, and that, the one connected to the main building. (Glover 1910, 10)

On the East coast at Ellis Island the greater range and number of those being processed meant that the facilities became spaces of research and experimentation as well as regulation. Many US and foreign scientists interested in the study of 'feeblemindedness', insanity and other newly defined pathologies travelled to New York to gaze upon the expert standards of classification being developed. People entering the island were subjected to an increasingly elaborate spectrum of tests which looked for pathologies as diverse as trachoma (by flipping up the eyelid with a button hook), illiteracy and imbecility (through doctor-designed paper and pen tests) and physical infirmities (by observing a person's gait and movement as they walked up a flight of steep stairs). Once 'diagnosed' arrivals would be inscribed with a symbol in chalk (each letter indicating a different pathology) to help inspectors further down the line make final decisions about entry. Migrants often negotiated this system in an expert manner, building up practiced knowledge about what various officials wanted to see and hear. Migrants would also hide the pathologies inscribed on their clothes by removing coats or turning them inside out.

As in Angel Island and other immigration stations operating in the late nineteenth and early twentieth century, the material construction of Ellis Island was designed to observe, channel, differentiate, and where necessary confine and imprison. The presence of medical isolation wards, a psychopathic ward, and caged exercise pens show that the material cultures of migration literally carved themselves into the landscape.

Commemorating Mobility

The link between mobility as an abstract indicator of legitimacy and mobility as a materialized spatial practice occurs in different, but equally significant, ways at these immigration stations today. Re-configured around the requirements of memory, Ellis Island and Angel Island have been appropriated by various institutions (the National Park Service, State Park bodies, National Trust for Historic Preservation, amongst others) to make abstract claims about a shared historical experience of mobility. The transformation in function of these islands from regulation to commemoration reflects a broader reconfiguration of the meaning given to the movement of those late nineteenth and early twentieth century immigrants originally processed. This in turn shapes both the physical orientation of the spaces and the way in which people move through them. It is the resultant material changes and their associated on-site social practices that interest us here. At Ellis Island ongoing discussions about the construction of a bridge to mainland New Jersey, the scheduling of ferries from Battery Park, the installation of elevators and escalators to ease passage around the museum, and the de-cluttering of interpretive spaces through a removal of cages, pens, fences, fixtures and fittings are all part of the *new* mobilities generated for an effective commemorative programme. Similarly, at Angel Island, the re-creation of an extant wharf, the excavation of original footpaths and walkways, the tacit dispersal of visitors away from the main resource and the construction compromises installed for compliance with the Americans with Disability Act are innovations that direct visitors' movement in a mnemonic enterprise.

After the destruction of the main administration building, the immigration authorities moved their operations off Angel Island leaving the remaining facilities to deteriorate until funds were given for their protection in the 1970s. Since then the barracks building which housed detained immigrants has functioned as a house museum interpreted by the Angel Island Association: a non-profit organization whose volunteers ran tours through the Immigration Station as part of a programme to interpret the island's many historic and natural resources. In 1997 the accreditation of National Historic Landmark status to the Immigration Station site transferred leadership of interpretation efforts to the California Department of Parks and Recreation, the National Park Service and the Angel Island Immigration Station Foundation. The station is currently in the middle of a thirty million dollar project entailing restoration and re-orientation where buildings are to be repaired, features replaced and visitor movement around the site adjusted. Instigating particular kinds of visitor movement is regarded as central to communicating the site's immigration history effectively. There are two main initiatives planned: the first concerns 'concentration', the second, 'diffusion'. Both are geared towards having visitors mimic the movements of immigrants so they might 'encounter and confront issues experientially in a way that enables them to feel compassion and responsibility' (Angel Island Immigration Station Visioning Workshops Report 1997, 7).

Efforts towards 'concentration' entail focusing movement around the 15-acre Immigration Station grounds rather than the island as a whole. This is to be achieved by the reconstruction of a 120-foot wharf that served the Immigration Station before being destroyed by fire. Currently, access to the site entails a 40-minute ferry ride to the main pier at Ayala Cove on the opposite side of the island followed by a one mile walk, cycle, or tram ride along a perimeter road that enters the grounds at the rear. This protracted journey limits the amount of time spent by visitors and school groups on tight schedules and gives a poor first impression of the resource. A dedicated wharf is deemed essential as a primary access point and interpretive entrance. It is also the point from which a programme of re-enactment can begin. Education and planning documents suggest an array of techniques to choreograph the movements of visitors in ways that imitate the on-site movements of immigrants as they were processed.

> The idea is to put the visitor in the position of the immigrant. Upon disembarking, the visitor could experience the sense of arrival and anticipation comparable to the immigrant who arrived here before … The simulated immigrant experience would continue as visitors approach the palm flanked Administration Building footprint. Re-enactment of the processing of immigrants and dreaded interrogation of Chinese Americans could occur on the footprint of the large Administration building. (Angel Island Immigration Station Visioning Workshops Report 1997, 9)

There is also a motivation for diffusion involving steering visitors away from the central resource of the barracks building. Narrative Treatment Guidelines provided by an Olmsted Cultural Landscape Report (Davidson and Meier 2002) make recommendations that include selective landscaping, clearing and planting to open up the site for visitors to explore the original circuitous pathways. In the process, visitors would apprehend ancillary functions relating to the processing and confinement of immigrants (a power house, hospital, workers cottages, and perimeter fencing) through a series of interpretive boards presenting the perspectives of officials, maintenance staff, and their resident families. Ultimately, the visitors' diffuse consumption of the grounds allows them to be co-opted in its animation. As interpretive architect and planner Daniel Quan notes, 'I really want to try and get people walking all over and seeing the whole thing as the station as opposed to just a bunch of buildings' (Quan 29 June 2002). Rather than an abandoned overgrown and forgotten location, carefully selected restoration would help portray the Immigration Station as peopled, dynamic and more emotive. Reflecting in more depth Quan notes:

> So I think that selectively if you take certain kinds of things – create an entrance that way and some other selective things perhaps where people walk. Right now your tour is just through the barracks, that first floor of the barracks and nothing else, but if you were allowed to walk outside into the recreation yard, which was fenced, and know that there wasn't any other way out, that you had to go back in you know, there are certain ways that you can do that to make people go through an experience where they see themselves enclosed inside of the fence. It hits home a little bit more. (Quan 13 October 2001)

The very construction of the Immigration Station at Angel Island as a space of memory works to populate the site giving it the appearance of a functioning immigration station and to construct a vicarious experience of immigrant processing by having visitors mirror migrants' on-site movements, bodily postures, and routines.

Ellis Island's restoration as a National Monument has entailed a similar change in the mobilities of visitors. A range of new opportunities for, and limitations on, movement have been created to help the space function more effectively for recollection. The selective demolition of less aesthetically pleasing buildings evidences very particular intentions regarding interpretation. The metal pens constructed to compress queues of migrants awaiting processing in the facility's Great Hall have been replaced by large sepia-toned photographs hung from the ceiling to keep floor space free. Stairs that immigrants were once forced to climb are still in place but visitors now also have the option of using an escalator or elevator to reach the upper floors. Knowledge of immigration at Ellis Island is presented through staged dramas of migration, uprootedness and alienation where visitors constitute 'recreational migrants' whose journeys to the island are driven by a desire to re-enact the immigrant experience albeit with a much shorter boat trip. The recasting of migrant history has been highly political and contested. National Park Service historians are now portraying the past in light of broader academic developments within heritage and museum studies. An expanded constituency of stakeholders has resulted in a polysemic message where migrants take on a range of characteristics: as subjects of honor inscribed by name on the 'wall of honor', as statistics visualized in 3D graphs and pie charts, as exotic subjects wearing foreign dress, and as imagined others whose disembodied voices crackle through the audio tapes of headphones.

The mobility of the site's employees has also been radically altered by restoration efforts, not least through the construction of a 400-yard land bridge from Liberty State Park in New Jersey to expedite renovation and reduce costs of the eventual 150 million dollar restoration project. While the bridge is in full view of citizens using Jersey City's bayfront promenade, a security booth restricts all those pedestrians not engaged in park official business (Low, Taplin and Shield 2005). The uneven politics of mobility surrounding the island's commemoration became starkly visible with coordinated opposition against New Jersey Senator Frank Lautenberg's proposal to purchase the bridge for public access. Historic preservation groups, the Circle Line Ferry Company, the New York Mayor's Office, the National Park Service, the Municipal Arts Society and the New York Times all opposed and ultimately defeated the plan with arguments as much about preserving for the visitor a 'true feel' of what immigrants experienced upon arriving by water (Myers 1991) as protecting the business interests of boat operations and services located at Battery Park in Manhattan. In a *New York Times* article Peg Breen, President of the New York Landmarks Conservancy, went so far as to comment 'the footbridge is a Trojan horse … There are some places in America that should stay special, that don't need highways' (MacFarquhar 3 April, 1997).

Now firmly embedded within the national heritage landscape, the interpretive programmes at Angel Island and Ellis Island do much more than disseminate competing narratives about their role in the regulation of early twentieth Century immigration. The curatorial agenda challenges our understanding of immigration as dichotomous – as either heroic or transgressive. Ellis Island's curators have been at pains to accentuate the experience of marginalized immigrants to temper the celebratory messages drawn out by visitors (see Maddern and Desforges 2004); those working at Angel Island seek to reach out in the opposite direction with attempts to align the uncomfortable history of Chinese exclusion to more appealing narratives about the peopling of America in general (see Hoskins 2004). In this way, curators and visitors alike struggle with the same theoretical tensions that surround mobility as social scientists.

Ambivalent Futures of Mobility

Limerick tells us that migration plays a central role in the creation myth of the United States – 'a tale explaining where its members came from and why they are special, chosen by providence for a special destiny' (Limerick 1987, 322). The contemporary resonance of these immigrant stories means that they are routinely appropriated by politicians presenting their visions of America's future. Ellis Island, in particular, is regularly used as a global public stage on which figures including Hillary Clinton, Rudolf Giuliani and President George W. Bush have, with the help of the media, delivered speeches about heritage, immigration control and foreign policy. The symbolism accrued at these immigration stations makes them key spaces upon which to articulate messages that draw legitimacy from an association with the immigrants of the past, whether it is George W. Bush addressing the world from the Great Hall at Ellis Island on the anniversary of September 11th conflating, in his speech, the 'war on terror' with the patriotic ideals of freedom and hope that drew millions of immigrants to New York harbour, or local officials presiding over citizenship ceremonies on Angel Island, where 'difference' is accepted and simultaneously instrumentalized to justify increased policing of the border.

In this chapter we have demonstrated how the on-site mobility experienced by people at these immigration stations has qualitatively transformed since their reconfiguration from spaces of regulation to spaces of commemoration. Key here is to appreciate that the spaces themselves play an important role in the construction of various types of mobilities practiced and experienced. As spaces of regulation Ellis Island and Angel Island informed precise discursive distinctions between national citizens and aliens ineligible for entry. As spaces of commemoration, Ellis Island and Angel Island codify that history of mobility as full of promise, as something to celebrate, something that unites and establishes a national identity. In the process, new mobilities arise in the consumption of these sites; mobilities used by managers, interpretive designers, and politicians in ways that make their messages more persuasive.

References

Agnew, J. and Smith, J. (eds) (2003), *American Space/American Place* (Edinburgh: Edinburgh University Press).

Angel Island Immigration Station Visioning Workshops Report (1999) (Berkeley: MIG Inc.).

Auge, M. (1995), *Non Place: Introduction to an Anthropology of Supermodernity*, translated by J. Howe (London: Verso).

Bamford, M. (1917), *Angel Island: The Ellis Island of the West* (Chicago: Women's American Baptist Home Mission Society).

Barde, R. and Bobonis, G. (2006), 'Detention at Angel Island First Empirical Evidence' *Social Science History* 30(1): 103–36.

Baudrillard, J. (1988), *America* (London: Verso).

Blunt, A. (2007), 'Cultural Geographies of Migration: mobility, transnationality and diaspora', *Progress in Human Geography* 31(5): 684–94.

Boorstin, D. (1966), *The Americans: The National Experience* (London: Weidenfeld and Nicolson).

Bourdieu, P. (1977), *Outline of a Theory of Practice* (Cambridge: Cambridge University Press).

Calavita, K. (2000), 'The Paradoxes of Race, Class, Identity and 'Passing': Enforcing the Chinese Exclusion Acts, 1882–1910', *Law and Social Inquiry* 25, 1–40.

Clifford, J. (1997), *Routes: Travel and Translation in the Later Twentieth Century* (Cambridge: Harvard University Press).

Crang, M. (2002), 'Between places: producing hubs, flows, and networks' *Environment and Planning A* 34, 569–74.

Cresswell, T. (2001), 'The Production of Mobilities' *New Formations* 43, 11–25.

Cresswell, T. (2003), 'Theorising Place' in T. Cresswell and G. Verstraete (eds) *Mobilizing Place, Placing Mobility* (New York: Rodopi Editions), 11–29.

Cresswell, T (2006) *On the Move* (London: Routledge).

Daniels, R. (1997), 'No lamps were lit for them: Angel Island and the historiography of Asian American immigration', *Journal of American Ethnic History* 71(1): 4–18.

Davidson, M. and Meier, L. (2002), *Olmsted Cultural Landscape Report for Angel Island Immigration Station* (Brookline, MA: Olmstead Center for Landscape Preservation).

de Certeau, M. (1984), *The Practice of Everyday Life* (Berkeley: University of California Press).

Gillis, J. (2003), 'Taking history off-shore: Atlantic islands in European minds' in Edmond, R and Smith, V. (eds) *Islands in History and Representation* (New York: Routledge).

Gillis, J. (2004), *Islands of the Mind: How the Human Imagination Created the Atlantic World* (New York: Palgrave Macmillan).

Glover, M.W. (1910), Letter to the Acting Commissioner of Immigration in San Francisco 21 November [Letter] History of Angel Island Immigration Area, California Department of Parks and Recreation, Sacramento, 10.

Honig, B. (2001), *Democracy and the Foreigner* (New Jersey: Princeton University Press).

Hoskins, G. (2004), 'A place to remember: scaling the walls of Angel Island immigration station', *Journal of Historical Geography* 30(4): 685–700.

Hoskins, G. (2005), *Memory and Mobility: Representing Chinese Exclusion at Angel Island Immigration Station* (Unpublished PhD thesis, University of Wales, Aberystwyth).

Kingston, M.H. (1980), *Chinamen* (New York: Ballantine Books).

Kouwenhoven, J. (1961), *The Beer Can by the Highway* (Baltimore: Johns Hopkins University Press).

Kraut, A. (1994), *Silent Travelers: Germs, Genes, and the Immigrant Menace* (Baltimore: Johns Hopkins University Press).

Kunstler, J.H. (1994), *The Geography of Nowhere: The Rise and Decline of America's Man-Made Landscape* (New York: Free Press).

Lee, E. (2003), *At America's Gates: Chinese Immigration during the Exclusion Era* (Chapel Hill: University of North Carolina Press).

Limerick, P.N. (1987), *Legacy of Conquest: The Unbroken Past of the American West* (New York: W.W. Norton).

Limerick, P.N. (1992), 'Disorientation and Reorientation: The American Landscape Discovered from the West', *The Journal of American History*, 79(3): 1021–49.

Low, S., Taplin, D. and Scheld. S. (2005a), *Rethinking Urban Parks* (Austin: University of Texas Press).

Low, S., Taplin, D. and Scheld, S. (2005b), The Ellis Island Bridge Proposal: Cultural Values, Park Access, and Economics, in S. Low, D. Taplin, S. Sheld (eds) *Rethinking Urban Parks* (Austin: University of Texas Press), 69–100.

MacFarquar, N. (1997), 'For Ellis Island, New Talk of a Hotel, a Bridge and Masses Yearning to Get In Free' *New York Times Online*, [internet], 3 April. Available at http://query.nytimes.com/gst/fullpage.html?res=9805E6D6143D F930A35757C0A961958260&scp=1&sq=ellis+island+bridge+Breen&st=nyt [Accessed 12 August 2008].

Maddern, J. (2005), *Spaces of History and Identity at Ellis Island Immigration Museum* (Unpublished PhD thesis, University of Wales, Aberystwyth).

Maddern, J. and Desforges, L. (2004), 'Front Doors to Freedom, Portal to the Past: History at the Ellis Island Immigration Museum, New York', *Social and Cultural Geography* 5(3): 437–57.

Markel, H. and Stern, A. (1999), 'Which Face? Whose Nation? Immigration, Public Heath, and the Construction of Disease at America's Ports and Borders, 1891–1928' *American Behavioural Scientist* 42(9): 1314–31.

Martin, A. (1997), 'The Practice of Identity and an Irish Sense of Place', *Gender, Place and Culture* 4(1): 89–119.

Myers, B.L. (1991), 'The Bridge to Ellis Island', *New York Times Online,* [internet], 3 February. Available at http://query.nytimes.com/gst/fullpage. html?res=9D0CE7D71E30F930A35751C0A967958260&n=Top/Reference/ Times%20Topics/Subjects/H/Historic%20Buildings%20and%20Sites [Accessed 12 August 2008].

Ngai, M. (1999), 'The architecture of race in American immigration law: A re-examination of the Immigration Act of 1924', *Journal of American History* 86(1): 67–92.

North, H.H. (1910) 'Report to the Commissioner General' 27 January [report] History of Angel Island Immigration Area, California Department of Parks and Recreation, Sacramento, 10.

Pelger-Gordon, A. (2006), 'Chinese Exclusion, Photography, and the Development of U.S. Immigration Policy', *American Quarterly* 58(1): 51–77.

Quan, D. (2001), Discussion on Immigration Station plans. [interview] (Personal Communication, 13 October).

Quan, D. (2002), Discussion on Immigration Station plans. [interview] (Personal Communication, 29 June).

Shah, N. (2001), *Contagious Divides: Epidemics and Race in San Francisco's Chinatown* (Berkeley: University of California Press).

Sheller, M. and Urry, J. (2006), 'The New Mobilities Paradigm', *Environment and Planning A*, 38, 207–26.

The Statue of Liberty-Ellis Island Foundation Inc. [website], (updated 22 Dec. 2008) http://www.ellisisland.org/genealogy/ellis_island.asp

Takaki, R. (1989), Strangers from a Different Shore: A History of Asian Americans (New York: Penguin Books).

Tuan, Y.F. (1974), *Topophilia: A Study of Environmental Perception, Attitudes, and Values* (New Jersey: Prentice-Hall).

Turner, F.J. (1947), *The Frontier in American History* (New York: Holt).

Tyner, J. (2006), *Oriental Bodies: Discourse and Discipline in U.S. Immigration Policy, 1875–1942* (Oxford: Lexington Books).

Wallenberg, C. (1988), 'Immigration through the port of San Francisco' in Stolarik, M (ed.), *Forgotten Doors: The Other ports of Entry to the United States* (London: The Balch Institute Press), 143–55.

Zelinski, W. (1973), *The Cultural Geography of the United States* (New Jersey: Prentice Hall).

Chapter 11
Cities: Moving, Plugging In, Floating, Dissolving

David Pinder

In many ways the essence of the city is the supreme coming
together for evrything [sic]
of it all
people come and go
it's all moving
the bits and pieces that form the city – they're expendable
it's all come-go

(Peter Cook 1963, 83)

The old fixed and static elements that built our cities are becoming increasingly irrelevant ... In a transient society, the mobile searchlight pinpointing an automobile sale or a movie premiere is more important than any building; a credit card system more meaningful than a high-rise bank. Urbanism, if it is to mean anything at all, is a fluid matrix of things that do their own thing. In William Burroughs' words, *we must keep our bags packed and ready to move all the time.*

(Warren Chalk 1969, 376)

Introduction

Cities moving. Cities raised on telescopic legs, venturing across territories and borders. Cities constructed as vast mobile machines, roving through deserts, oceans or urban terrains. From where do these strange urban voyagers come? To where are they heading? Gathering in the waters off Manhattan, are they arriving or leaving (Figure 11.1)? Is their intent invasion, salvation or communication? Does their movement indicate new freedoms made possible by a technologically advanced age? Or does it have more threatening or disturbing connotations, suggestive perhaps of militaristic marauding or of the evacuation of the island in the wake of disaster? In these now iconic images of post-war avant-garde architecture from Ron Herron's Walking City from 1964, cities are literally given legs. They are depicted singularly or in clusters as they walk and meet, connect and dock. At times, drawn flat in elevation, they take on the appearance of giant insects. At others they appear more like the mutant offspring of existing industrial

Figure 11.1   Ron Herron, Cities Moving: New York, 1964 © 2010 by Ron Herron, courtesy of the Ron Herron Archive.

and military constructions, such as the vast space rocket assembly units found at Cape Canaveral, Florida, or the anti-aircraft installations at Shivering Sands Fort, with their interconnected towers perched on legs in the Thames Estuary.

The project arose from Herron's involvement with the English Archigram group that, during the 1960s, shook up the architecture establishment through provocative images, texts, exhibitions and projects. Walking City, which appeared in *Archigram* 5, is one of the group's most outlandish and best-known images. It also most obviously embodies the group's deep concern with mobility. It was one of many projects produced either individually or collaboratively between 1961 and 1974 by the group's loosely connected six core members – Warren Chalk, Peter Cook, Dennis Crompton, David Greene, Ron Herron and Michael Webb – that embodied a shift away from the static, rooted and monumental towards movement, flexibility, transitoriness and indeterminacy. It was also one of a range of architectural, artistic and urban projects by modernist and avant-garde groups in Europe, North America and beyond during the 1960s that challenged fixed urban structures and experimented with mobility and mobile spaces. At stake was not simply an emphasis on movement and speed, a commonplace in modern planning visions concerned with improving the pace and efficiency of transit. Rather, there was a fundamental questioning of conventional ideas about cities and buildings as fixed objects in the effort to leave behind static forms, to embrace movement and flexibility, even to give them legs or flight.

This chapter addresses such calls for cities to become mobile. My focus is on selected urban and architectural projects from the European avant-garde of the late 1950s and 1960s. Their attempts to embody a restlessness, flexibility and nomadism that they held to be in keeping with modern times may at times seem fantastical yet exploring aspects of their mobile turn is significant not only for histories of modernist architecture and urbanism, in relation to which they have recently gained increasing attention, but also for thinking critically about influential conceptions of cities and mobilities more generally. Urbanism, note Mimi Sheller and John Urry (2006, 2), 'has always been associated with mobilities and their control, and continues to be so more than ever. The technologies, infrastructure, material fabric and representational machinery of cities support these mobilities, while also being shaped and re-shaped by them.' Among the ways in which understandings of cities have been increasingly mobilised in recent years has been through critical studies of networked infrastructure, such as transport, telecommunications, water, energy and the like. These have provided powerful perspectives on the dynamic processes that constitute modern urbanism, allowing cities to be seen as 'staging posts in the perpetual flux of infrastructurally mediated flow, movement and exchange', and highlighting their roles 'in articulating the corporeal movements of people and their bodies' (Graham and Marvin 2001, 8; see also Amin and Thrift 2002). Mobilities have also come to the fore through multidisciplinary studies concerned with how city spaces and structures are used, practiced and performed (for example, Borden et al. 2000; Latham and McCormack 2004; Merriman, this volume). Another illuminating lens through which to consider these themes, I want to suggest, and in particular to reflect on their long-standing as well as politically charged and contested nature, can come from reconsidering urban visions from the recent past.

Among the questions to be asked are: how were the avant-garde proposals for mobile cities imbued with critical and emancipatory intent? How did they figure cities and their inhabitants as nomadic, and to what effects? How did they address the long-standing tension in urban thought between movement and settlement (Mumford 1961; Allen et al. 1999)? Of particular concern in what follows is how the protagonists connected mobility with freedom, and how they consequently struggled against dominant concerns with the regulation and management of urban movement. This entailed confronting the spatial orderings of conventional modern urbanism and especially its institution of means through which materials and bodies are kept in their 'proper' conduit and place. At the same time, however, there is a need to question the social and political ramifications of their consequent celebration of flow, flux and mobility. Tim Cresswell refers to the latter generally as a 'nomadic metaphysics' and notes its prominence not only in contemporary architecture theory practice but also in current cultural and intellectual life, where 'it is beginning to appear everywhere' (2006, 51). Through developing historical geographical perspectives on earlier nomadic conceptions of cities, I want to consider what critical light might be cast on such tendencies today?

Mobilities for the Atomic Age

When Archigram and the other urbanists and architects to be discussed in this chapter proposed their mobile cities during the 1960s, they wrestled with contemporary approaches to urbanism and architecture, and with what they saw as an overbearing and now stodgy influence of officially sanctioned modernism.

Existing urban and architectural forms inhibited progressive change and stifled creativity, so many critics contended. They exerted a crushing weight that helped to support outdated social and urban practices. Modernist urban visions may have once promised radically different futures but they now seemed increasingly absorbed by bureaucratic states. As embodied by the Congrès Internationaux d'Architecture Moderne (CIAM), the modern movement was meeting increasing internal criticism, including from the likes of the Independent Group and Team 10. Influential schemes such as CIAM's Athens Charter, drawn up in 1933 and published ten years later, and based on making clear separations between the functions of residence, work, recreation and traffic, were being criticised for being too rigid and closed in their attempts to order and regulate changes, and hence as unable to respond to the demands of a new era. Large-scale urban planning projects were also being confronted by political opposition from residents demanding that their voices be heard.

The appearance of Archigram's magazine in the early 1960s posed an exhilarating challenge to conventional urban construction and architectural practice. In the words of one of its members it was initially 'an outburst against the crap going up in London' and 'against the attitude of a continuing European tradition of well-mannered but gutless architecture that had absorbed the label "modern", but had betrayed most of the philosophies of the earliest "modern"' (Cook 1967, 133). Against ideals of permanence, its members embraced the transient, the ephemeral and the expendable qualities of modern urban existence. Against austere and pared down forms, they proposed exuberantly colourful schemes that explored the potential of new technologies for maximising pleasure and fun. Against the separation of functions and specialisms, they broke down barriers between disciplines and artistic spheres, between low and high culture. And instead of behaving themselves within fixed ideal plans, elements of their collaged urban visions were found swinging from cranes, clipping on, plugging in, lifting off, inflating, hovering, zooming, floating with multi-hued balloons, and being projected onto temporary surfaces. Far from being opposed to modernism, they sought to recapture the earlier experimental and oppositional energies of the futurists and constructivists while also drawing on the forces being unleashed by the pace of capitalist modernisation, technological development and the acceleration of consumerism and travel as Britain emerged from post-war austerity. They were, as they put it, 'in pursuit of an idea, a new vernacular, something to stand alongside the space capsules, computers and throw-away packages of an atomic/electronic age' (Archigram 1963, 112). In a spirit more playful and joyful than angry they took their imagery as readily from science fiction, popular magazines and comics as from technological hardware and space-age constructions, and they disseminated the results through exhibitions, lectures, publications and a journal with an urgency indicated by their name, derived from ARCHItecture teleGRAM.

Archigram had considerable international influence, especially on architecture students and those looking for alternatives to modernist orthodoxies that were closer to the spirit of swinging sixties London and the wider counter-culture, an influence that has increasingly been recognised recently within architectural history with the staging of exhibitions, belated establishment recognition in the form of a Gold Medal from the Royal Institute of British Architects in 2002, and the appearance of the first full length historical studies (Sadler 2005a; Steiner 2009). Among early projects was the Plug-In City, an ambitious programme designed by Peter Cook in 1962–1964 and driven by his question: 'what happens if the whole urban environment can be programmed and structured for change'? (Cook 1964, 33). His proposals built on discussions of expendability. They broke with ideas of architecture as a play of static forms and scrambled the functional demarcation of zones and activities that was enshrined in the Athens Charter. Besides the organised framework, which he envisaged as potentially extendable across the channel from Britain into other parts of Europe, everything was flexible. Units and capsules could be plugged in or removed through a system of cranes, while services and transportation were provided by tubes, pipes, monorails, hovercraft and the like. Principles of transience and expendability were foregrounded, with the projected obsolescence of components varying from around forty years for the main structure to a few years for rooms. The plug-in concept owed much to Archigram's positive appraisal of urban life and vibrant crowds conveyed in their Living City exhibition of 1963, in which they investigated the city's 'movement-cycles', and in relation to which they argued: 'Cities should generate, reflect, and activate life, their environment organised to precipitate life and movement. Situation – the happenings within spaces in the city, the transient throwaway objects, the passing presence of cars and people – is as important, possibly more important, than the built demarcation of space' (Archigram 1963, 112).

If urban megastructures were designed with flexible and movable components, so too were houses. Archigram questioned the assumed psychological need for a house in the form of a static and permanent container. Commissioned in 1967 to design a future house for the year 1990, they regretted the stipulation that it should be in a fixed location but nevertheless made the walls, ceilings and floors adjustable. The robot-serviced interior included inflatable seating and sleeping arrangements, and a living room chair that doubled as a 'chair-car' based on hovercraft principles to allow excursions into the city beyond (see Archigram 1994). Pneumatic structures more generally allowed experimentation with amorphous, transient, portable and almost immaterial environments, a theme that would also be explored by numerous other individuals and groups that included Utopie, Haus-Rucker-Co and Coop Himmelb(l)au. Archigram developed its own interest in mobile cities and spaces through an array of stations, capsules and pods through which architecture could be mobilised and individuals could source public services while having the freedom to roam. As David Greene (1994 [1966], 182) put it in relation to his Living Pod, 'With apologies to the master, the house is an appliance for carrying with you, the city is a machine for plugging into'. More recently he has suggested that, in retrospect, Archigram projects provide 'a new agenda where nomadism is the dominant social force; where time, exchange and metamorphosis replace stasis; where consumption, lifestyle and transience become the programme; and where the public realm is an electronic surface enclosing the globe' (Greene 1998a, 3).

The nomadic theme and the group's 'longtime devotion to the notion of motion', as Michael Webb (1998, 98) put it, was taken further as Archigram developed environments that could be carried or worn. Functioning as portable environments for explorers and wanderers, Webb's inflatable Cushicle (1966) and Suitaloon (1967) and Greene's Inflatable Suit-Home (1968) came complete with television, water supply, food and heating. In Herron's related Nomad (1968), an action-man-like figure appeared with an environment kit venturing far from urban terrain. Archigram thus pushed towards the dematerialisation of architecture in moves that had parallels at the time in the art world. It sought to go 'beyond architecture', in its own phrase from *Archigram* 7, or at least to redefine it so that 'architecture would become more like a refrigerator, car, or even a plastic bag than an immovable monolith' (Sadler 2005a, 107). Cities were figured more in terms of images, information, events and the electronic equipment needed to create them than anything resembling traditional buildings. Such is apparent in proposals for an Instant City (1968–70), which could be assembled from place to place and so temporarily bring the intensities of urban living to different communities. In this regard Archigram members found compelling the huge temporary gatherings soon to be seen at rock festivals such as Woodstock in 1969. Cities and buildings even rescinded from view completely in Greene's Bottery that same year, in which citizens wandered in a wired garden, plugging in their portable TVs and other hardware into conveniently located Rokplugs or Logplugs, which blended into the surrounds as part of a 'fully serviced natural landscape'. 'Modern nomads need sophisticated servicing', stated Greene, 'and in the Bottery this is achieved by the technique of calling it up wherever you are, it's delivered by robots' (Greene 1998b [1967], 144, 151). He portrayed this as an architecture related to time, devoid of formal aesthetic statements, meant to disturb the environment as little as possible: a kind of 'invisible guerrilla environment' (159). Outlining the possibilities of a Local Available World Unseen Network (LAWUN), he claimed that the implications for the mobilisation of cities in a virtual sense would be immense: 'The whole of London and New York will be available in the world's leafy hollows, deserts and flowered meadows' (Greene 1969).

Designing for Unsettlement

Archigram's emphasis on mobility and flexibility can be seen as part of a wider interest in these themes among modernist and avant-garde architects and urbanists across Europe and beyond during the 1960s. Social, economic, technological and demographic changes were demanding new ways of thinking about cities and urban life, so many believed. The static categories of conventional urban planning and architecture were no longer adequate, it was claimed, and for some the very idea of permanent settlements had been undermined and even rendered obsolete in an increasingly mobile age. Richard Buckminster Fuller, whose early experiments with electronic networks and spatial structures had made him a guru-figure in advanced debates about architecture and technology during the period, referred in 1963 to living 'under conditions of mobility which result in continual stirring up rather than settling down' (cited in Wigley 2001, 122, n. 76). Unlike many critics or planners who decried this disturbance, and who thus sought to fix and restore settlements, he believed the role of designers lay in 'accommodating human unsettlement', with technological means fostering new kinds of mobility through new world networks. As early as 1927 he projected a 'one world town' in which everything was mobile and in which physical infrastructure was replaced by 'atomized nomadic systems' (Wigley 2001, 113).

The emphasis on mobility and flexibility was for many architects connected with a desire to question their role as authoritative form givers intent on producing permanent structures according to the rules of fine design, and to turn instead to mobile structures and open frames where control of internal form might be increasingly relinquished to users. A common theme was that of 'open ends', a term Simon Sadler (2000) takes from an editorial in *Archigram* 8 to characterise a wider shift in the modernist paradigm during the 1960s away from the idea that buildings and plans should be conclusive and expressions of clear programmes. Megastructures were frequently projected as the basis for such openness with the advent of new materials and technologies at the time, along with techniques for space frames pioneered by such figures as Fuller and Konrad Wachsmann, making the construction of lightweight, mobile and even floating structures seem increasingly feasible. Archigram's magazine served as an important means of communication and provocation in carrying images from a range of pioneering projects beyond their own, its fifth issue in particular featuring work by Constant, Yona Friedman, Eckhard Schulze-Fielitz, Paolo Soleri, Karel Tange and many others alongside Herron's Walking City.

How each of these figures approached urban mobilities varied considerably, despite the illusion created of a kind of 'Megastructure International' by Archigram's magazine (Banham 1976, 89) and by many subsequent exhibitions and publications. Constant's New Babylon project, for example, initiated while he was a member of the Situationist International during the late 1950s, and then pursued independently throughout the following decade through art works, writings, lectures and other means, displayed similar interests to those taken up by Archigram in technological developments, in increasing mobility and leisure, in the potential automation of work, and in the ability of space frames to enable new freedoms. But, in sharp contrast, it was based on an implacable opposition to capitalism and contemporary urbanism, and its realisation explicitly depended upon revolutionary social and spatial transformation. Constant resisted the idea that it was an urbanistic project as he sought to envisage a post-revolutionary world in which people were liberated to shape their own spaces according to their activities and desires, and in which a new nomadic and ludic way of life resulted from being freed from work-based constraints (see Wigley 1998; Pinder 2005;

Sadler 2005b). Alongside Archigram, however, I want to set the work of another influential figure who promoted a mobile architecture and urbanism during the 1960s in an attempt to respond to the challenges of the time and to expand the freedoms of urban inhabitants, and whose work provides a significant comparison and contrast: namely, the Hungarian-born architect Yona Friedman.

Friedman traces his point of departure back to a single question that occupied him as a student: 'why should architects decide for the people who live in their buildings?' He believed that 'people should decide for themselves', a principle that should extend from personal dwellings to public spaces (2005a, 30). His interest in mobility arose from this desire to maximise individual liberty within cities while ensuring their environments could meet present needs as well as those to come. Having been disappointed with CIAM discussions on adaptability and mobility in Dubrovnik in 1956, Friedman issued the first version of his *L'Architecture mobile* the following year when he also moved to Paris, and he founded the Groupe d'Etudes d'Architecture Mobile (GEAM) the year after. Friedman and his fellow members attributed many urban problems to the rigidity of urban fabric, to its inability to be adapted for life as it is lived. They saw their schemes as a response to current demands that included those associated with population growth, technological development, increasing leisure time, traffic congestion, housing shortages, the imprisoning nature of the built environment, and above all the belief that people should be able to determine their own environments according to the moment. Through an 'indeterminate town planning', as Friedman termed it, inhabitants would be provided with flexible and interchangeable structures so that they could adapt and re-make their spaces as required. The stated aim of his group was to 'render the problem of static form outmoded', and among their proposals were those for movable walls, floors and ceilings, easily alterable infrastructure networks, and large mobile spatial units that would supposedly be like cities themselves, taking the form of 'interchangeable containers (travelling, flying, floating)', 'buildings on rafts' and 'buildings bridging over spaces' (GEAM 1970 [1960], 168).

Friedman elaborated his programmes for a mobile architecture and urbanism over many years, both through GEAM and well beyond it after the group's demise in 1962. At their centre was the distinction also important in Archigram megastructural schemes between an advanced infrastructure, the principles of which he sought to outline, and its flexible contents, which he left open to the erratic nature of reality so that it could be arranged and rearranged by individuals according to need (Friedman 1970a [1959], 1993; Vlissinge and Lebesque 1999). With this in mind Friedman proposed gigantic space-frames, raised from the ground on pilotis that opened the third dimension of the grid to moving, dwelling and working in all directions. He depicted these both as abstract forms and as raised above existing cities whose old cores, he contended, would be left intact (Figure 11.2). Through this multi-layered grid and a process of 'superposition', he aimed to reintegrate functions whose spatial segregation was at the heart of the Athens Charter. And by insisting that components within the rigid structure should be demountable, movable and transformable, he contended that this approach would not imprison further growth and change but would be able to meet the unpredictable needs of the future while simultaneously allowing individual initiative and liberty to thrive. Inhabitants would register their individual preferences for living spaces through what he termed a Flatwriter, and suitable spatial arrangements would then be calculated according to available possibilities. Friedman's rejection of permanence and his advocacy of flexibility extended to institutional and organisational norms. He argued, for example, that property rights should be subject to renegotiation every ten years, and marriage every five years.

Friedman was a leading figure among a group of 'spatial urbanists' in France that proposed utopian urban designs during the 1960s based on the possibilities they believed were opening up through technological developments. The diversity of their positions and their complex entanglements with each other as well as with international colleagues and contacts preclude brief summary. As Larry Busbea (2007) has recently discussed, however, many shared Friedman's interest in the potential of urban constructions floating above the ground as well as concerns with the potentially liberating impacts of new technologies and automation in taking care of material needs, in extending leisure time, in enhancing the circulation of

Figure 11.2. Yona Friedman, Spatial City, project. Perspective, 1958-59. Felt-tipped pen on tracing paper. 34.3 x 49.5cm. New York, Museum of Modern Art (MoMA). Gift of The Howard Gilman Foundation. Digital image © 2010, The Museum of Modern Art, New York/Scala, Florence

residents through urban spaces and in enabling new forms of nomadism. While their proposals often appeared fantastical, they also at times engaged with specific urban problems and administrative realities. In his schemes for Spatial Paris, for example, Friedman portrayed his structures raised above the city centre as a means of tackling current issues of growth and traffic congestion. He argued that they could effectively add spaces for housing, business, industry and even agriculture, as much as tripling housing densities, but at the same time preserve the existing city below. They would also facilitate the sorting of traffic by assigning pedestrians and automobiles their own spaces, one in the floating thoroughfares and the other in existing arteries on the surface (Friedman 2005b).

Of Nomads and Nomadic Spaces

As has already been stressed, there are important differences between Archigram and Friedman in their concerns with urban mobility, not to mention between them and other theorists and architects from the same period with whom they are often linked. What particularly interests me here, however, is the value that many of these projects gave mobility as part of a wider interest in the experiences and practices of users, in space as becoming and performed, and in the event of architecture as something realised by inhabitants themselves. Mobility, in keeping with influential ideologies of modernity, was connected with liberty, opportunity and choice. It was something to be extended to all residents against top-down direction and control exerted by planners. To emphasise mobile urbanism in this way seemingly involved siding with individual agents against prescription and closed forms. Everyone could be an architect or builder, so it was implied, in the sense that everyone could be enabled to adjust and rearrange her or his own environments through the provision of suitable equipment and the dissolution of architecture as traditionally known. Behind such visions was an optimistic view of technological development as enabling new freedoms and a belief that the role of the urban planner and architect was to facilitate such freedoms rather than to impede them with fixed constructions and schemes. However implausible some of the assumptions about moving into a global post-scarcity society might appear today, it was a view that protagonists claimed was based on actual social, economic and technological conditions rather than fanciful projections into the future.

In the process the projects provided compelling perspectives on mobility in opposition to the authoritarianism of much modernist urbanism that seeks to fix and frame movement within an ideal form. In particular they connected mobility to issues of freedom, play, unpredictability and happenstance. As they developed these themes, many avant-gardists of the 1960s turned to the figure of the nomad. David Greene directly evoked a 'Cowboy international nomad hero' in relation to their efforts to devise portable environment kits, as someone who was 'probably one of the most successful carriers of his own environment' (1998b [1967], 151). References to nomads and nomadism more generally abounded in work by Archigram and by the spatial urbanists in France as well as by many other architects and artists at the time to grapple with emerging urban conditions and possibilities, and to explore potential uses of mobile, portable and pneumatic structures, either prospective or historical. Their appearance came at a time of growing concern with international migration, displacement and travel, in relation to which the nomadic could be viewed not simply as an aspiration but as a common if differentially experienced condition, one that had as much to do with upheaval, uprooting, dislocation and feelings of homelessness associated with deterritorialisation as with the ability to roam freely by choice or as a leisure activity (Busbea 2007, 55–6; see also Scott 2001). Yet in evoking nomads in this context, the avant-gardists conjured a figure that was critical in relation to sedentarising powers of state planning and positive about the prospect of finding new modes of using space within conditions of deterritorialisation, of riding new flows and currents rather than retreating from them. They often thus invested the term nomad with liberatory connotations, as Greene's reference to the cowboy hero suggests.

Such moves reclaimed the nomad from its negative treatment by many nineteenth and twentieth century western urban critics, planners and architects who not only often associated it pejoratively with a past and primitive state, one that was supposedly superceded with the construction of towns that marked a higher order of existence, but also depicted it as threatening, disordered, uncontrolled, the outcome of the transgression of boundaries and borders (see Cresswell 1997; Sibley 1995). They may be compared with the reversal of negative discourses about the 'primitive' nomad in more recent social and cultural theory that has explored its transgressive and subversive potential, and that has been particularly influenced by Gilles Deleuze and Félix Guattari's (1987) discussions of nomadology. Yet this reversal has also entailed risks and problems. Notable among them has been a romanticisation of the nomad whereby an encounter with an imagined non-European other, in distinction to actual nomadic peoples, enables a conceptualisation of urban mobility that remains unlocated, ungrounded, disembodied and 'unmarked' (Cresswell 2006, 53).

The nomad that wandered unbounded through Archigram's graphics and writings specifically was invariably male and white, as the reference to cowboys suggests, and it is not hard to view much of the space-age imagery and gadgetry that so fascinated the group as predominantly male fantasies, of a particular class and location at that. This is perhaps not surprising given the group's deliberate immersion in popular commercial imagery of the time as well as their all male participation in a male-dominated profession, about which the choice of consumer items in Chalk's Living City Survival Kit of 1963, with its arrangement of cigarettes, dark glasses, jazz records, convenience foods, bottle of whisky, copy of *Playboy* and replica gun, among other accoutrements of an urban wanderer, offers wry commentary (Sadler 2005a, 72–7). But the differential abilities of bodies to move, the different relationships that different social groups have to mobility, and the geographically and historically specific ways in which mobility is socially produced were insufficiently examined. Remaining obscure were the power geometries along lines of class, gender, ethnicity and disability that enable certain kinds of mobility and that might enable alternatives. Understanding them is crucial for appreciating differences between, for example, the movements of those forced to migrate after displacement from homes, neighbourhoods or countries and those free to initiate journeys and flows, something that Tim Cresswell and Janet Wolff among others have critically addressed in relation to celebratory discourses of the nomad and of travel more generally within cultural theory and criticism (Cresswell 1997, 2006; Wolff 1993). There is nothing inherently liberating in mobility, as these writers have stressed, for it needs to be understood in terms of particular power relations and settings, as something social and spatial rather than originating from within individual agents. This brings into question the politics of the avant-garde's promotion of nomadism and urban mobility, to be discussed further in the next section.

Freedom and Control

For all the remarkable ways in which Archigram and Friedman in particular presented mobile cities that challenged certain dominant conceptions of urban design and space, they left unaddressed fundamental issues about power and about the social, political and economic processes through which mobilities and urban spaces are produced. In Archigram's case the valorisation of free movement and circulation was couched in terms of individual choice and accompanied by the assertion that such choices and nomadic behaviours were becoming available to all, not simply the wealthy. 'We are not politically over-developed as a group,' admitted Archigram's Peter Cook (1967, 133), 'but there is a kind of central emancipatory drive behind most of our schemes'. Archigram's emphasis was on removing constraints and obstacles while providing technological support for a nomadic life. A key question was: 'do buildings help towards emancipation of the people within? Or do they hinder because they solidify the way of life preferred by the architect?' (Archigram 1968, n.p.). In this regard they believed architecture was currently failing, asserting: 'If only we could get to an architecture that really responded to human wish as it occurred then we would be getting somewhere'.

Archigram's route was to plunge into consumer and popular culture, to seek out their potentially liberating force. Their projects embraced the logic of capitalism and they readily presented them as consumer products: 'the real justification of consumer products,' they stated in the article quoted above, 'is that they are the direct expression of a freedom to choose' (Archigram 1968, n.p.; see also Cook 1967). Many of their capsules, pods, and kits were designed with the comforts and pleasures of individual clients and consumers in mind. 'Choice means freedom', 'What you want when you want', 'What you want where you want it', ran slogans from the graphic 'Metamorphosis' that was part of their Control and Choice project in 1967. Such slogans were pitched against what Archigram saw as the austerity, dullness and moralising attitudes of the British architectural scene at the time, as its members drew inspiration from the commercial culture and open freeways of the United States, where several of them came to settle (Fraser and Kerr 2007, 295–6). By binding their technologically-fuelled anti-authoritarianism so closely to ideologies of consumer choice and individual freedom, however, they were unable to gain critical perspectives on how desires and needs are shaped rather than simply met under capitalism, on how mobilities are differentially structured along axes of power, or on how their rhetoric of choice and freedom feeds into advocacies of the free market. Their approach certainly owed much to contemporary counter-cultural and radical political practices, and they looked forwards to an 'anarchic city'. At the same time they presented themselves as apolitical rather than committed and they came under fire from both the left and right. Hence the assessment that in many respects their libertarian individualism and their attitudes to deregulation conforms to 'the neo-liberal ethos of late capitalism' (Curtis 1998, 61).

In contrast to Archigram's festive and playful schemes, Friedman's drawings of spatial structures like those of many fellow spatial urbanists were stark and almost empty. In Friedman's case this was in keeping with his insistence that 'fitting out of the skeletons [of the infrastructure] will depend upon the initiative of each inhabitant' (1970b [1962], 183), and his decision 'to look at the minimum departure, trying to leave the page as blank as possible' (2005a, 32). Consumer culture was kept at arm's length although Friedman referred favourably to commercial producers who encouraged the participation of customers to adapt and alter products, and suggested that architects and urbanists could learn from their example (Friedman 1970a [1959]; Busbea 2007, 73). More fundamentally there was again little attempt to question underlying social, economic and political processes. The freedom of inhabitants appeared to rest upon promises of technological progress coupled with the provision of a flexible environment that prevented constraints on movement and change, rather than on more thoroughgoing social and political change. Inequalities of power around class, gender and ethnicity went unaddressed. Indeed, they were apparently superseded in images of the free flow of inhabitants, whose frictionless movement was not in fact untethered but rather held stable within the giant frames. Inhabitants were thus able to rearrange the furniture, so to speak, but not to change the framework, which was lifted into the air and ideally meant to be scarcely visible.

The technocratic cast of the schemes became particularly apparent when presented in proposals for cities such as Paris. In attempting to integrate with existing urban conditions rather than to confront the underlying causes of their problems and injustices, the shallow notion of freedom at their heart was exposed. As Busbea comments, in such projects the promises of mobility were narrowly circumscribed and channelled into a vision of the city as a circulatory system, the main concern of which was less to forge a democratic and emancipatory urbanism than to manage 'circulatory equilibrium for a congested Paris' (2007, 135–6). In this respect it repackaged more than challenged visions familiar from earlier modernist urbanism with its demands for circulation and efficient flow. Caught between 'cybernetic fantasy and administrative reality' during the 1960s, so Busbea suggests, spatial urbanists more generally raised their profile by working with existing governments and their priorities but 'essentially tied their own critical fate to that of mainstream modernism' (136). Others have argued that, for all the claims of transferring power to residents through leaving everything open, the proposals for flexible structures can be interpreted as expanding rather than reducing opportunities for technocratic control. As Felicity Scott puts it, 'the structure simply provided a more elaborate illusion of freedom into which the subject could be integrated' (2001, 237 n. 67).

Revolutionising Cities and Mobilities

The need to develop critical perspectives on visions of mobile cities from the 1960s was recognised by many contemporary critics. Not surprisingly Archigram's schemes in particular disturbed establishment figures, among them prominent members of the modern movement such as Sigfried Gideon and Constantinos Doxiadis. But they also attracted the ire of many radicals who viewed them as technological fetishists giddy on the speed up of capitalist gadgetry. When Ron Herron presented his Walking City at an Archigram organised conference in Folkestone, in 1966, he was met with cries of 'war machine' and accusations of fascism and totalitarianism from some of the predominantly student audience (Jencks 1985, 292; Sadler 2005a, 155). Assertions that it represented a destructive force seem to have surprised Herron himself, who professed to have always seen it as 'a rather friendly-looking machine' that 'moved slowly across the earth like a giant hovercraft, only using its legs as a leveling device when it settled on its site' (1994, 75). More generally, by the late 1960s, questions were increasingly being asked, not least by architects themselves, about the meaningfulness of flexibility and 'choice' within megastructural and other technological schemes when the options were merely among those set by the designers. Also under scrutiny were the new kinds of controls that they entailed, particularly through the roles of the 'programmer' and 'manager', in relation to which the construction the following decade of the Centre Pompidou in Paris, as a state-backed megastructure, provided a telling case (Banham 1976, 204–16).

The Marxist philosopher Henri Lefebvre contemporaneously asserted radical perspectives on mobile cities in the French context. He disparaged the proliferation of spatial schemes devised by urbanists and architects at the time, and insisted that 'the social and professional mobility so desired by planners (primarily urban planners and moving companies) is fundamentally superficial' in that it refers to 'the displacement of populations or materials that leave social relationships intact' (2003 [1970], 97). He was particularly concerned about their programmed and structured approach to people's movements, which became a means of managing the contradictions and tensions of capitalist urbanisation in favour of 'equilibrium' and 'stability'. This entailed tightening rather than relinquishing controls and constraints through forms of urban planning that imposed homogeneity and suppressed play, he argued, and to that end he opposed the machinations of technocrats that he saw as the enemies of cities in extending commodification and abstract spaces. With the work of spatial urbanists in France in mind, he wrote:

> What of the residential nomadism that invokes the splendours of the ephemeral? It merely represents an extreme form, utopian in its own way of individualism. The ephemeral would be reduced to switching boxes (inhabiting). To suggest, as Friedman does, that we can be liberated through nomadism through the presence of a habitat in the pure state, created through metal supports and corrugated steel (a giant erector set), is ridiculous. (Lefebvre 2003 [1970], 98)

While Lefebvre's charges of technocracy applied most strongly in the French case, his arguments about the desire for mobility and associated claims of freedom being insufficient unless tied to a critique of current power relations have wider purchase, including in relation to Archigram's attempts to go beyond megastructures and even architecture in favour of serviced landscapes. For urban change to be liberating, Lefebvre stressed, it must involve the transformation of both everyday life and space. It requires the appropriation of urban spaces by inhabitants so that cities are no longer alienated products or commodities but *oeuvres* that are consciously and collectively created by inhabitants. It involves the claiming of rights to the city, the assertion of use values over exchange values. The primary agents would be social movements and groups that 'would invent their moments and their actions, their spaces and times, their works. And they would do so at the level of habiting or by starting out from that level (without remaining there; that is, by modelling an appropriate urban space)' (2003 [1970], 99).

In this regard, Lefebvre was drawn in particular by Constant's New Babylon as an attempt to envisage a mobile urbanism based on a revolutionary transformation of space and society. Lefebvre's interest was based on his appreciation of the radical political differences between Constant and Friedman and the spatial urbanists. These differences have too often since been elided as their work has frequently been exhibited and published alongside one another, yet they came out clearly in dialogues between them. In correspondence with Friedman, Constant (1961) accused him of failing to extend his critique beyond technical and practical dimensions to address the social and cultural transformation of cities. Constant believed that Friedman overemphasised private dwellings and crucially underestimated questions about collective creativity in relation to games, everyday life and space. More generally Constant refused to design for current social conditions in a way that might reinforce those conditions, something he criticised Friedman for doing. Instead he projected New Babylon towards an other city and an other life, believing that the functional city based on production could give way to cities and urban life based upon collective creativity, play and free wandering (Constant 1996 [1974]; for further discussion, see Pinder in press).

Conclusions

The mobilities of the roaming, plugging-in, floating and dissolving cities of the 1960s avant-garde have typically been presented as liberating but, as this chapter has discussed, their politics were at best ambivalent and need interrogating. Important questions arise through contemporaneous work by Lefebvre among others, who emphasised that liberation cannot be ensured by spatial design or technological development alone. Nor can it be reduced to smoothing the paths for individual movement, however innovative the forms proposed, for it requires a transformation of fundamental social and spatial relationships, including those that constitute the basis of capitalist society and its unequal power relations. The politics of space entails struggles and conflicts over the processes through which space is produced. More than movement per se, what is crucial politically is attaining the right and power to move or, equally importantly, to remain in place. The latter may indeed at times be the critical concern for people as their landscapes evaporate through processes of capitalist creative destruction, and as their land and resources are seized and populations displaced through rounds of accumulation by dispossession, hence the significance of assertions of 'the right to stay put' (Hartman 1984) as a component of 'the right to the city'.

These arguments have particular importance today when, rather than simply contesting dominant norms and controls, an emphasis on mobility, flexibility and continual change might be as easily portrayed as complicit with the demands of neo-liberal capitalist urbanisation and the injustices that it wreaks. For does it not feed into and even bolster capital's self-narration as hyper-mobile, anticipating and sketching out a fluid world in which, in the words of Marx and Engels, 'all that is solid melts into air'? Might not visions of mobile cities in this way function pedagogically to prepare the way for and to normalise aspects of capitalist urbanisation and its creative destruction? In the absence of social and economic change, it is not difficult to see the images of walking cities and infinitely flexible space frames discussed above as emblematic more of the free movement of capital than that of the liberation of people. Troubling questions similarly shadow the floating and dissolving architectures addressed elsewhere in this chapter, many of which gloss over or spirit away social divisions along lines of class and other axes of power, and as a consequence could be seen as heralding the 'universal fluidity [that] is the new ideology of structurelessness in a postmodern cultural era' (Smith 2008, 154).

At the same time, however, reconsidering their remarkable visions of mobile cities casts important light on ideas and practices of urbanism, and on contested conceptions of mobility, which are too often taken for granted or reduced to technical matters. They also remain striking for the assumptions, paradoxes and contradictions they reveal, among them those fundamental to the urban process under capitalism. As David Harvey has long demonstrated in his writings on this subject, investment in the built environment and associated spatial structures for transportation is necessary for the accumulation of capital, and yet the immobility of those structures and the sinking of capital into place comes to hinder the drive to expand, to overcome spatial barriers and to find new opportunities for profitable investment that is central to capitalism's operation and survival. New urban landscapes are produced in the image of capital appropriate to its needs at a particular time but their physical forms also constrain and imprison that process and they subsequently require devaluation and replacement. Capitalist development is thus characterised by continuous struggles over the creation and destruction of physical landscapes (Harvey 1985, 24–5). From this perspective, the visions of urban mobility and flexibility considered in this chapter might seem like the materialisation of a capitalist fantasy or as having, at the very least, an ambivalent relationship to capitalism's own expansionary and nomadic drives. But then they also raise another question: how might the freedoms of movement that they present so vividly be claimed by people? How might control over movement be taken by urban inhabitants, not as a purely technological or technocratic concern, but as one involving the fundamental basis on which cities are determined as social spaces?

Acknowledgements

Earlier versions of this chapter were presented at Columbia University, at the University of California Los Angeles, at Manchester Metropolitan University, and at the Annual Meeting of the Association of American Geographers in San Francisco. Many thanks to Tim Cresswell and Pete Merriman for originally inviting the essay and for comments; to audiences at those events for engaging with the materials; and to James Connolly, Lisa Kim Davis, Tim Edensor, Peter Marcuse and Justin Steil among others for their hospitality.

References

Allen, J. Massey, D. and Pryke, M. (eds) (1999), *Unsettling Cities* (London and New York: Routledge).
Amin, A. and Thrift, N. (2002), *Cities: Reimagining the Urban.* (Cambridge, UK: Polity).
Archigram (1963), 'Situation', in T. Crosby and J. Bodley (eds), 112. Reprinted in Archigram (ed.), 88.
Archigram (1968), 'Open Ends', *Archigram* 8, n.p. Reprinted in *Archigram* (ed.), 216–26.
Archigram (1994) 'Living 1990', in *Archigram* (ed.), 197–9.
Archigram (ed.) (1994), *A Guide to Archigram 1961–74* (London: Academy Editions).
Banham, R. (1976), *Megastructure: Urban Futures of the Recent Past* (London: Thames and Hudson).

Borden, I., Kerr, J., Rendell, J. with Pivaro, A. (eds) (2000), *The Unknown City: Contesting Architecture and Social Space* (Cambridge, MA: MIT Press).

Busbea, L. (2007), *Topologies: The Urban Utopia in France, 1960–1970.* (Cambridge, MA: MIT Press).

Chalk, W. (1969), 'Things that Do their Own Thing', *Architectural Design*, 376.

Conrads, U. (ed.) (1970), *Programs and Manifestoes on 20th-Century Architecture.* (Cambridge, MA: MIT Press).

Constant (1961), Letter to Yona Friedman, dated 21 April, in Constant Archive, Rijksbureau voor Kunsthistorische Documentatie (RKD), The Hague.

Constant (1996 [1974]), 'New Babylon', in L. Andreotti and X. Costa (eds), *Theory of the Dérive and Other Situationist Writings on the City* (Barcelona: Museu d'Art Contemporani de Barcelona/Actar), 154–69.

Cook, P. (1963), 'Come-go: the key to the vitality of the city', in T. Crosby and J. Bodley (eds), 80–3.

Cook, P. (1964), 'Plug-In City', *Sunday Times Colour Magazine*, 20 September, 33.

Cook, P. (1967), 'Some Notes on the Archigram Syndrome', *Perspecta* 11, *Amazing Archigram: A Supplement*, 133–7.

Cresswell, T. (1997), 'Imagining the Nomad: Mobility and the Postmodern Primitive', in G. Benko and U. Strohmayer (eds), *Space and Social Theory: Interpreting Modernity and Postmodernity* (Oxford: Blackwell), 360–79.

Cresswell, T. (2006), *On the Move: Mobility in the Modern Western World* (New York and London: Routledge).

Crompton, D. (ed.) (1998), *Concerning Archigram* (London: Archigram Archives).

Crosby, T. and Bodley, J. (eds) (1963), *Living Arts* 2 (London: Institute of Contemporary Arts and Tillotsons).

Curtis, B. (1998), 'A Necessary Irritant', in D. Crompton (ed.), 25–79.

Deleuze, G. and Guattari, F. (1987 [1980]), *A Thousand Plateaus: Capitalism and Schizophrenia Volume 2*, trans. B. Massumi (Minneapolis: University of Minnesota Press).

Fraser, M. with Kerr, J. (2007), *Architecture and the 'Special Relationship': The American Influence on Post-War British Architecture* (London and New York: Routledge).

Friedman, Y. (1970a [1959]), *L'Architecture mobile: vers une cité conçue par ses habitants* (Paris: Casterman).

Friedman, Y. (1970b [1962]), 'The Ten Principles of Space Town Planning', in U. Conrads (ed.), 183–4.

Friedman, Y. (1993 [1959]), 'Program of Mobile Urbanism', in J. Ockman with E. Eigen (eds), *Architecture Culture 1943–1968: A Documentary Anthology.* (New York: Rizzoli), 274–5.

Friedman, Y. (2005a), 'In the Air: Interview with M. van Schaik', 28 October 2001, in M. van Schaik and O. Mácel (eds), 30–5.

Friedman, Y. (2005b [1961]), 'Paris Spatial: A Suggestion', trans. P. Aston, in M. van Schaik and O. Mácel (eds), 18–29.

GEAM (1970 [1960]), 'Programme for a Mobile Architecture', trans. M. Bullock, in U. Conrads (ed.), 167–8.

Graham, S. and Marvin, S. (2001), *Splintering Urbanism: Networked Infrastructures, Technological Mobilities and the Urban Condition* (London and New York: Routledge).

Greene, D. (1994 [1966]), 'Living Pod,' in Archigram (ed.), 182.

Greene, D. (1994 [1969]), 'Children's Primer', *Architectural Design*, May. Reprinted in Archigram (ed.), 297.

Greene, D. (1998a), 'A Prologue', in D. Crompton (ed.), 1–4.

Greene, D. (1998b [1967]), 'The World's Last Hardware Event', in D. Crompton (ed.), 144–59.

Hartman, C. (1984), 'The Right to Stay Put', in C. Geisler and F. Popper (eds), *Land Reform, American Style* (Totowa, NJ: Rowman and Allanheld), 302–18.

Harvey, D. (1985), *The Urbanization of Capital: Studies in the History and Theory of Capitalist Urbanization* (Oxford: Blackwell).

Herron, R. (1994), in R. Banham (ed.), *The Visions of Ron Herron* (London: Academy Editions).

Jencks, C. (1985), *Modern Movements in Architecture*. (Harmondsworth: Penguin).

Latham, A. and McCormack, D. (2004), 'Moving Cities: Rethinking the Materialities of Urban Geographies', *Progress in Human Geography*, 28(6): 701–24.

Lebesque, S. and van Vlissingen, H.F. (eds) (1999), *Yona Friedman: Structures Serving the Unpredictable*. (Rotterdam: NAi Publishers).

Lefebvre, H. (2003 [1970]), *The Urban Revolution*, trans. R. Bononno. (Minneapolis: University of Minnesota Press).

Mumford, L. (1961), *The City in History: Its Origins, its Transformations and its Prospects* (London: Secker and Warburg).

Pinder, D. (2005), *Visions of the City: Utopianism, Power and Politics in Twentieth-Century Urbanism* (Edinburgh: Edinburgh University Press).

Pinder, D. (in press), 'Nomadic Cities', in G. Bridge and S. Watson (eds) *A Companion to the City* (Oxford: Blackwell, new edition).

Sadler, S. (2000), 'Open Ends: The Social Visions of 1960s Non-Planning', in J. Hughes and S. Sadler (eds) *Non-Plan: Essays on Freedom, Participation and Change in Modern Architecture* (London: Architectural Press).

Sadler, S. (2005a), *Archigram: Architecture Without Architecture*. (Cambridge, MA: MIT Press).

Sadler, S. (2005b), 'New Babylon versus Plug-In City', in M. van Schaik and O. Mácel (eds), 57–67.

Scott, F. (2001), 'Bernard Rudofsky: Allegories of Nomadism and Dwelling', in S. Williams Goldhagen and R. Legault (eds), *Anxious Modernisms:*

Experimentation in Postwar Architectural Culture (Cambridge, MA: MIT Press), 215–37.

Sheller, M. and J. Urry (2006), 'Introduction: Mobile Cities, Urban Mobilities', in M. Sheller and J. Urry (eds) *Mobile Technologies of the City* (London and New York: Routledge), 1–17.

Sibley, D. (1995), *Geographies of Exclusion: Society and Difference in the West* (London and New York: Routledge).

Smith, N. (2008), 'Book review essay: Castree, N. and Gregory, D (eds), David Harvey: A Critical Reader', *Progress in Human Geography* 32(1): 147–55.

Steiner, H. (2009), *Beyond Archigram: The Structure of Circulation* (London and New York: Routledge).

Van Schaik, M. and Mácel, O. (eds) (2005), *Exit Utopia: Architectural Provocations 1956–1976* (Delft: IHAAV-TV and Munich: Prestel).

Webb, M. (1998), 'The Notion of Motion', in D. Crompton (ed.), 98–105.

Wigley, M. (ed.) (1998), *Constant's New Babylon: The Hyper-Architecture of Desire.* (Rotterdam: Witte de With/010 Publishers).

Wigley, M. (2001), 'Network Fever', *Grey Room* 4, 82–122.

Wolff, J. (1993), 'On the Road Again: Metaphors of Travel in Cultural Criticism,' *Cultural Studies* 7(2): 224–39.

PART III
Subjects

Chapter 12

Commuter:
Mobility, Rhythm and Commuting

Tim Edensor

Rhythm, Place and Mobility

While the recent turn to mobility has been of enormous benefit in interrogating static versions of place and space, there has been little discussion about the relationship between mobility and temporality. However, Henri Lefebvre's hitherto neglected *Rhythmanalysis* (2004) offers a starting point for investigating the complex temporal rhythms of the multiple mobilities that course through space. This chapter thus tentatively examines how rhythmanalysis might be applied to commuting while also critiquing academic and popular cultural discourses that represent commuting as a dystopian, alienating practice. In popular cultural representations, the commuter has often been represented as a frustrated, passive and bored figure, patiently suffering the anomic tedium of the monotonous or disrupted journey (for instance, as in the 1980s British television sitcom, *The Rise and Fall of Reginald Perrin* where the main character suffers the daily banality of underground 'tube' commuting and in the opening scenes of Hollywood movie *Falling Down*, where the angry protagonist leaves his car in the midst of a cacophonous traffic jam). In addition, as Lyons et al. (2007, 108) claim, many studies of commuting continue to construe travel-time as a 'disutility', a cost 'incurred by individuals and society as a means to enjoy the benefits of what is available at the destinations of journeys'. Instead of these dystopian and functional visions, and within a broader analysis of the relationalities between rhythm, mobility and space, I explore the spatial and experiential dimensions of commuting rhythms and argue that commuting can be alternatively conceived as a mobile practice that offers a rich variety of pleasures and frustrations.

Commuting belongs to an enormous range of spatio-temporal contexts within which multiple rhythms are produced and interweave. Social rhythms may be institutionalised (marked by national and religious occasions, hours of commerce or television schedules) or produced through synchronised habits (eating, playing and working together). Rhythms may be linear or cyclical and operate at circadian, weekly, monthly, seasonal, annual, lifetime, millennial and geological scales. They can be regular or irregular and vary according to the time and space between events, tempo and intensity, degrees of predictability and disruption, and the coinciding effects that produce polyrhythmicality, synchronicity or dissonance. Importantly, the subjective experience of the rhythmic and temporal may vary so that time can drag or appear to be going too quickly.

Lefebvre's *Rhythmanalysis* (2004) is something of an unfinished project and though it contains certain theoretical consistencies and numerous insightful and provocative passages, it should be conceived as introducing a rich but suggestive vein of temporal thinking rather than a definitive methodology or a set body of concepts, which opens up a fertile realm of exploration, a starting point for thinking about rhythm. And while Crang appositely critiques Lefebvre's ideas for foreclosing the productive potential for becoming through time (as opposed to repetition emphasised by Lefebvre's emphasis on repetition and the discreteness of the present moment), I regard the project as sufficiently open to adapt and expand notions of rhythm (also see Edensor and Holloway 2008). More specifically, I now identify three key benefits of thinking about the rhythms of mobility.

First of all, rhythmanalysis elucidates how places possess no essence but are ceaselessly (re)constituted out of their connections. For instance, cities are particularly dense spatial formations containing a complex mix of multiple, heterogeneous social interactions, materialities, mobilities and imaginaries which connect through 'twists and fluxes of interrelation' (Amin and Thrift 2002, 30), are 'economically, politically and culturally produced through multiple networked mobilities of capital, persons, objects, signs and information' (Urry 2006, ix). Places are thus continually (re)produced through the mobile flows which course through and around them, bringing together ephemeral, contingent and relatively stable arrangements of people, energy and matter. A rhythmanalysis of commuting might 'open up all sites, places, and materialities to the mobilities that are always already coursing through them' and demonstrate how '(P)laces are thus not so much fixed as implicated within complex networks by which hosts, guests, buildings, objects, and machines are contingently brought together to produce certain performances in certain places at certain times' (Sheller and Urry 2006, 209). Rhythmanalysis thus emphasises the dynamic and processual qualities of place, circumventing overarching sedentarist spatial reifications in contradistinction to 'time geographies' which abstractly spatialise or map time onto place. Spaces are thus always immanent, in process, fecund and decaying. In this context, a travelling human is one mobile element in a seething space pulsing with intersecting trajectories and temporalities, and does not occupy a vantage point from which space can be known. As Lefebvre says, '(There is) nothing inert in the *world*', which he illustrates with the examples of the seemingly quiescent garden that is suffused with the polyrhythms of 'trees, flowers, birds and insects' (Lefebvre 2004, 17) and the forest, which 'moves in innumerable ways: the combined movements of the soil, the earth, the sun. Or the movements of the molecules and atoms that compose it' (ibid., 20).

The organised braiding of multiple mobile rhythms produces distinct forms of spatio-temporal order, maintained through the orchestration of traffic management systems, the conventions of travel practice, the affordances of transport forms and the characteristics of the space moved through. A rhythmanalysis can identify these rhythms through which spatial order is sustained (and thwarted or challenged). As Amin and Thrift argue, despite the multi-temporality of the city, it is rarely subject to manifold chaos. Rather, the synchronisation of thousands of rhythms, reiterative social practices ranging from official strictures to work and leisure rituals, shape routine urban experience, are the 'repetitions and regularities that become the tracks to negotiate urban life' (2002, 17). The diurnal pace of urban life varies within and between cities, with their hectic rush hours, quiet mid-afternoons, vibrant early evenings and low-key nights. Superimposed on this are the habits of individuals, their body rhythms, seasonal and 'natural' rhythms, and broader institutional rhythms of media and officialdom, so that place is constituted by a multitude of rhythmic combinations that 'fold time and space in all kinds of untoward localisations and intricate mixtures' (Amin and Thrift 2002, 47). In any space, the activities and interactions of numerous social actors intersect, as suburbanites, young, old, shopkeepers, dog-walkers, the religious and the festive, drug addicts, schoolchildren, shoppers, workers, traffic wardens and students collectively constitute space through their rhythmic and arhythmic practices. Commuting practices add to this rhythmic mosaic, co-ordinating the performance of some rhythms, enfolding and clashing with others. Places are thus not typified by any single rhythm or temporality but are locations at which multiple temporalities collide, synchronise and interweave (Crang 2001) to produce an ever changing polyrhythmic constellation, yet one which produces distinct spatialities. For instance, Degen (2008) highlights how the diurnal, weekly and annual rhythms of the regenerated heritage district of Castlefield, Manchester are constituted by the separate activities of groups of people drawn to specific areas and commercial events with little coincident polyrhythms, whereas by contrast, el Raval district in Barcelona is typified by the co-ordination of insider and outsider activity rhythms where different spatial practices constantly overlap in space and time, with no rhythm dominating.

In developing an understanding of rhythmic spatialities, I avoid the dominant tendency to focus upon the rhythmic flows that centre upon a particular place (for example, Crang 2001) and examine the ways in which rhythm shapes the mobile experience of space. Although I do not suggest that apprehensions of place are ever static, the speed, pace and periodicity of a journey produce particular effects through which space and place are known and felt, a stretched out, linear apprehension shaped by the form of a railway or road, the qualities of the vehicle and the time and pace of the journey. Where this is experienced through the regular, repetitive spatio-temporal trajectories of commuting – every weekday at the same time, and along the same route – a distinct embodied, material and sociable 'dwelling-in-motion' emerges (Sheller and Urry 2006). Place is thus experienced as the passing of familiar fixtures under the same and different conditions of travel, which produce a sequence of generally regular events and phases within a particular (clock-based) time frame.

Augé (1995) contends that the mobile production of space produces 'non-places', and Cresswell also notes how some geographers maintain that 'places marked by an abundance of mobility become *placeless*' (2006, 31) realms of detachment. However, such assertions overlook 'the complex habitations, practices of dwelling, embodied relations, material presences, placings and hybrid subjectivities associated with movement through such spaces' (Merriman 2004, 154). Indeed, as Jiron (2007) points out, in commuting, a sense of inhabitation inside a bus or tram is intermittently folded together with a sense of those places passed on the journey. Such place-making and sensing is rhythmic.

Secondly, foregrounding rhythm underscores how humans are 'rhythm-makers as much as place-makers' (Mels 2004, 3). Accordingly, a study of rhythms identifies the temporal organisation of space and forms of subjective temporal experience. Commuting sews places together and produces an itinerary shaped by time, as temporalities of movement are continually reinscribed on places and periods of travel en route. And the intimate rhythms produced in the interaction between passengers and vehicles inhere in the individual timetables through which people build up a relationship with place(s) over a span of time. On the move, they produce their own individual temporalities whilst ignoring or conforming to larger, collective scheduling patterns, thereby simultaneously contributing to the (changing) rhythms of place. The collective, simultaneous enaction of mobile rhythms constellates around particular places at particular times, for instance, with the typical two 'rush hours', in early morning and late afternoon. These rhythms add to the production of diurnal urban temporalities through various forms of mobility or its absence, including the lunchtime bustle, mid afternoon lull, early evening anticipation in the quest for entertainment, and late night quiescence.

Thirdly, this latter example highlights how an exploration of mobile rhythms can elucidate the relationship between imposed and individual rhythms. A key issue for Lefebvre is the tension between the corporeal, social and spatial regulation which keeps things ticking over according to normative temporal schema, and the arrhythmic episodes and inclinations which are apt to disrupt this timekeeping. He argues that '(P)olitical power knows how to utilise and manipulate time, dates, time-tables' (2004, 68). Through the use of clock-time to dominate everyday temporalities, power instantiates the regularisation of work and school hours. Officialdom also regulates the rhythms of mobility, attempting to ensure that mobile subjects conform to authorised tempo, speed, synchronicity and pacing. One effect of these regulations is that rhythms become habitually and unreflexively embodied by those who participate in their performance, and are difficult to knowingly contravene. Common sense notions of appropriate timings inhere in normative ways of understanding and experiencing the world, instantiating conventions about when and where particular practices should occur.

These official rhythms are also accompanied by bodily, seasonal and cosmic rhythms which are apt to disrupt them. Lefebvre points out that the 'rational, numerical, quantitative and qualitative rhythms superimpose themselves on the multiple natural rhythms of the body ... though not without changing them' (2004, 9). Yet though corporeal rhythms may become accommodated to temporal procedures, the body is also apt to find discord with delineated rhythms according to chance and event: 'to become insomniac, love-struck or bulimic is to enter into another everydayness' (Lefebvre 2004, 75). Similarly, mobile subjects are not always subservient to the functional requirements of traffic flow and management. Travellers appropriate time through creative activity whilst unexpected intrusions produce arrythmias and uneven rhythms. For rhythms continually change, intersecting and flowing in diverse ways, and where they are apparently repetitive and regular, they are apt to be punctured, disrupted or even curtailed by moments and periods of arhythmy. As with other regulatory practices, the production of rhythmic order, requires a remorseless monitoring that is impossible to maintain at all times (Graham and Thrift 2007).

Commuting involves the channelling, monitoring and scheduling of a privileged form of mobility, in contradistinction to the perceived threats of nomad and vagrant movement (Cresswell 2006), coercing commuters 'to orchestrate in complex and heterogeneous ways their mobilities' (Urry 2006, 19) within managed 'routescapes'. Commuting also intersects with other work, domestic, leisure and social routines of greater or lesser regularity as well as unplanned or improvised events. Accordingly, within this broader rhythmic pattern, 'the temporal organisation of a day can be hypothesised as one co-ordinated around fixed events that usually involve the co-participation of others' (Southerton 2006, 45). The co-ordination and synchronisation of routines are thus part of a bigger scenario of rhythmic planning which must accord with others' routines, established temporal strictures and contingencies and improvisations. For instance, Urry contends that automobility 'forces people to juggle tiny fragments of time' combining freedom and flexibility, and constriction (2006, 20). Such juggling must accommodate changes, as with the adaptations of commuters to the recent introduction of a congestion charge in London.

The Rhythms of Commuting

The rhythms of commuting are exceedingly diverse and shaped by numerous factors, including the mode of transport and its particular affordances, cultural practices and social conventions, modes of regulation, the distance travelled and the specificities of the space passed through. Commuters who travel on the New York subway, the Tokyo bullet train, or by car in Calcutta are enmeshed in distinct rhythms and experiences. Certain forms of mobility coincide, where streams of traffic coagulate as they near urban centres, whilst others remain quite discrete, such as driving along a sparsely populated linear route in an 'edge city'. The spatial scales of travel vary according to whether a journey conjoins home and work over short distances or extends many miles, or occurs on large highways, smaller roads or railways. And the temporal scale of travel similarly diverges between a ten minute drive and a two hour long journey, and there are variations in the number of stops en route, the evenness of the mobile flow, the numbers of bends and obstacles, the practices of co-commuters, and the affordances of the road or rail track. Such factors might produce staccato, fluid or wildly variable rhythms (see Spinney 2007, for a discussion of various cycling rhythms). Contemporaneously, commuting practices increasingly diversify as work hours become more varied and flexible, producing a plethora of synchronic and asynchronic mobile rhythms. Yet despite this multiplicity, certain prevalent commuting characteristics persist.

For commuting is a distinctive variant of circadian travel, primarily distinguished by its particular temporal context within routinised diurnal (Southerton 2006) and weekly patterns, having a roughly allotted time, and a position within a sequence of different activities in different places that organises the daily distribution of staying put and moving. Within this routine, there is a repetitive temporal span within which travellers must attempt to get to work or back home, predictable destinations, departure and arrival times, route consistencies and mobile rituals, and the establishment of meeting points to achieve co-presence (Kaufmann 2002). Accordingly, these regular patterns of 'social practices, coded gestures, metaphorical styles, technological applications and experiences' (Mels 2004, 6) extend forms of shared mobile experience.

The specificities of commuting can contrast with other mobile practices that are inflected with different cultural resonances, produce different rhythms and temporalities, and cover different spaces. For instance, the leisure-oriented weekend 'tootle' in the car, popular amongst British weekenders, foregrounds a contrasting *laissez faire* mobile disposition to destination and route, while the cross-continental journey by bus surrenders agency in making time and choosing a route. The trans-American road trip is a tootle stretched out over time and space whereas the short temporal punctuations of the supermarket visit or school-run prioritise the functional imperatives of dutiful travel. However, while it may appear that certain kinds of travel are suffused with meaning and pleasure by contrast with commuting, commuting time is not dead or neutral time that simply links more meaningful spatial contexts. For as Cresswell (2006) declares, the content of the line between A and B (or between work and home, and home and work) has remained rather under-explored.

I now discuss some distinctive rhythmic effects and experiences of commuting in order to open up mobilities to temporal analysis. I identify key themes of synchronicity, consistency, disruption, and sensation. Throughout the discussion, a tension emerges between the pleasures and irritation of predictability, and the delights and frustrations of disruption and improvisation, and this engenders a more realistic understanding of the pleasures and pains of commuting than that offered by dystopian representations. The passages will be leavened by extracts from an account of one young female commuter, Gemma:

> For the last 7.5 years I have been travelling between Crewe and Macclesfield to work and back. The experience varies on a daily basis; I drive for between 45 minutes and an hour and a half each way, dependant on traffic and weather conditions, the routes is about 23 miles but takes so long because it's cross country.

Commuting Synchronicities and their Disruption

As in other areas of routinised everyday life, various technologies of regulation – the maintenance of schedules, ticket collectors and other guardians of conduct, traffic rules and laws, traffic flow systems, warning notices, policing, signage, road markings and speed cameras – keep the driving commuter in check and maintain the consistency of certain codes of conduct. These are further enhanced by a move from 'the oligoptical surveillance and self-disciplining of drivers, to automated systems of management using…software-enabled and distributed technologies that mediate in various ways road infrastructure, vehicles, and drivers. (Dodge and Kitchin 2007, 265). Regulatory systems attempt to maintain the flow and speed of traffic and necessarily impose a regulated and synchronised rhythm upon commuters, with penalties for failure to conform. However, these laws and conventions are commonly flouted, notably where the smooth flow of movement is thwarted and drivers break rules about overtaking or exceed the legal speed limit. Most obviously, a more regulated synchronicity prevails in the shared experience of rail, bus and tram passengers which, along with commuters' internalisation of performative conventions and the collective enaction of habitual cultural norms, facilitates 'people together tackling the world around them with familiar manoeuvres' (Frykman and Löfgren 1996, 10–11). Such synchronicities are productive of social relations whether these include face-to-face interaction or not:

> Sometimes I just like to see other people though, it makes me realise I am not alone, there are lots of other people going about their business, it makes me smile and fill with a 'love' for all these crazy living beings on the planet and the highs and lows that each of them/us goes through on a daily basis. (Usually) I just drive past them (but) I bumped into one of my 'drive by' people recently and said 'hello'. They hadn't a clue who I was as obviously they had never seen me in my car.

Synchronicity also arises through the weave of vehicles going at the same pace, or where there is more than one lane, a limited series of different paced mobile rhythms that produce a collectively constituted rhythmic choreography. The stop-go rhythms and points on the journey where vehicles decelerate are part of this temporal choreography, which can be identified through the time of the journey or in place. Cronin (2006) has shown how regular traffic rhythms are identified by advertisers who place billboards at locations where travelling vehicles slow or stop. These visual commercials contribute to urban rhythms in their augmentation of particular stopping points, and in their marking of periodicity, through the length of their installation.

Travel is commonly an occasion for sonic immersion (Bull 2001), and besides the customisation of individual listening strategies, collective commuting soundscapes are produced by radio programming which develop 'drivetime' slots and regular traffic bulletins. Journeys can thus be shaped by media rhythms, with chat, news programmes and variously spaced musical interludes. Roads and railways are thus replete with commuters simultaneously listening to these programmes, routinely producing a synchronised rhythmic experience of mobility.

Yet although these routinised, synchronic rhythms are bureaucratically regulated and collectively produced, this does not necessarily determine the potential for commuters to tailor their journeys in accordance with their own strategies, imperatives and feelings:

> When I first started the journey in 1999 it was a long hard slog, with lots of concentration required through areas of the county that I didn't know well at all. I was stuck in a rut going the same way day in day out and not really taking in the beautiful countryside I was passing through. I felt very tired at that time and didn't get any enjoyment from what I classed as my torturous journey. It was eating my day – get up in the dark get home in the dark … I decided to make the change after about six months of this and started altering my route to and from work. This made the journey very exciting because I hadn't a clue where I was at times. It also opened up lots of new places to me that have captivated me since … I now change my journey daily; the route taken is usually on a whim and dependant on how tired I am, and my mood.

Moreover, the synchronic flow of mobile vehicles and people must be continually reproduced and is apt to be disrupted by asynchronic disruptions of all kinds – accidents, the dwindling capacities of road and rail infrastructures, failures in regulatory systems or adverse weather conditions – and also encounters with those moving at different rhythms. An example of this latter disruption occurred occasionally during my previous daily commute by car from Manchester to Stoke-on-Trent, where along the A556, the short link road between two motorways, the M56 and the M6, a herd of dairy cattle would block the road for five to ten minutes as they slowly made their way from field to farm, creating an enormous traffic jam.

Pleasure and Consistent Rhythms

Although frequently represented as dreary and alienating, the very consistency of the repetitive rhythms of commuting can, on the contrary, permit a diverse range of pleasurable effects. The daily apprehension of regular features provides a comforting reliability, fostering a mobile homeliness and a familiarity with space as particular sights mark the stages of the journey (Edensor 2003, 2008). Moreover, the phases of the journey reliably mark progress, including stages at which traffic speeds up or slows, and routinised elements such as the purchase of the daily newspaper or travel tickets which enfold social relations into the daily ritual. We might also consider the serial features that install a sense of spatial belonging, including the road signage and roadside furniture that occur with rhythmic regularity (for instance, consider the metronomic swish of regularly spaced telegraph poles passed at the same pace which act as recurrent visual sequences and musical notations). This rhythmic experiential consistency also includes sites recurrently passed that belong to a larger space – the familiar retail outlets, places of worship, institutions and architectural styles – that constitute a sense of being in national or regional space (Edensor 2004). Such cyclically recurrent scenes need not be conceived as tediously familiar but as consistent yet changing elements in a landscape.

> It seems that when I'm happy I drive the longer routes to work which take me through the ever changing countryside, sometimes passing Jodrell Bank. This is always a great route to take no matter what the time of year or weather as the 'dish' is always different. It can be in cloud, bright sunlight, mist, shining eeriness, dark and menacing … a chameleon … I sometimes wish I had a camera with me to capture the changing face of Jodrell Bank at these times, kind of a daily log in some respects just so people could experience the differences I see… the same goes for the fields, trees, animals and people I pass.

This underappreciated familiarity, the regular passing of familiar landmarks, people, events and objects at a predictable speed allows certain nuanced and enhanced appreciations of place through mobility to be grasped. For that which stands out from the norm is highlighted. The sudden road accident, newly painted house, unusual bird or animal, peculiar vehicle or passer-by: all stand out in sharp relief to the usual happenings, and the strange is especially enchanted by its occurrence within a normative realm. This familiar mobile spatial knowledge fosters a sense that space and places change, and processes of change – a new building under construction or the sowing of a new crop – can be seen from inception to completion. Such familiarity is further enriched by an appreciation of seasonal changes, such as the changing light, foliage and wildlife, inducing an apprehension of our enfolding within the larger temporal rhythms of a space of the changing same and the momentarily surprising. Paradoxically then, mobile experiences of place and belonging may be transient and fleeting as well as associated 'with prolonged or repeated movements, fixities, relations and dwellings' (Merriman 2004, 146).

The regular rhythms of mobility also facilitate a host of activities that can be undertaken when action has become automatic and unreflexive, where commuters are on 'auto-pilot, free and absorbed in the moment' (Ford and Brown 2006, 159). As commuting becomes sedimented as regular, the habitual forms of negotiating space – finding a bus seat or driving along a particular road – become procedures requiring little or intermittent attention. Like all habits, these provide, a way to economise on life, to let go of self-monitoring and reflexivity. In their extensive survey of rail passengers, Lyons et al. (2007, 110–13) reveal that 62 per cent of commuters read for leisure, half watched others or looked outside, a third worked, and smaller percentages snoozed, socialised and listened to music. Strikingly, they report that only two per cent of all respondents found travelling boring and less than a quarter of commuters considered that time spent commuting was wasted.

The automaticity commonly ascribed as boring, allows commuters to dream, read, telephone and plan, and being mobile is increasingly mediated by technologies such as in-car hi-fi systems, mobile phones and i-Pods which are woven into routines and produce habituated embodied interactions. For example, I previously drove 40 miles, often listening to the car radio as I commuted. Following its theft, I continually reached out to the dashboard to switch on the absent radio, an instance of embodied automacity that testified to the synchronised rhythm of body and car. The habitual inhabitation of space can also foster improvisational listening strategies and forms of social interaction which allow for continual modulation of the sonic environment:

> I can debate about a million and one topics each day just from random radio listening! The radio station I chose also changes with my mood. Radio One, Signal One, Radio Four, Rock FM, Classic FM. whatever catches me as I change the channel (and) ... sometimes I chat on my phone on the way home.

Rhythmic predictability thus offers opportunities for carrying out fulfilling, useful and valuable activities which may be constrained by a 'time squeeze' during more purposive periods. It is common for drivers to gain information about news or to listen to music en route to work and they may also engage in the less instrumental activities of talking gibberish, dancing whilst seated and face-pulling outside of normative social constraints.

> The solitude of the journey is a big positive, it gives me time to shout, scream, cry, laugh, sing at the top of my voice and be free with only my self to worry about and no chance of anyone from life interfering in this 'alone time' It also makes me realise that I don't spend a lot of time in general on my own, so it's a very important piece of my day now, even if it is in a car.

Moreover, within the routinised structure of the day, commuting, despite being part of a mobility that connotes restlessness and dynamism, can offer a period of relaxation, a temporal hiatus from more pressing imperatives. Or its taken for granted homeliness may provide opportunities for carrying out those duties for which there was no time at home, for instance, putting on make-up and having breakfast, as do the bus commuters in Santiago described by Jiron (2007).

Mobile Sensations and the Apprehension of Space and Time

Forms of transport possess different affordances and provoke diverse performances, and they are 'experienced through a combination of senses and sensed through multiple registers of motion and emotion' (Sheller and Urry 2006, 216). However, commuting has been represented as a deeply unsensual experience. For example, Sennett claims that the micro movements entailed in car-driving in general lighten the 'sensory weight' of the body, 'suspending the body in an ever more passive relation to its environment' (1994, 375) and he refers to 'the slight touch on the gas pedal and the brake, the flicking of the eyes to and from the rearview mirror (which) are micro notions compared to the arduous physical movements of driving a horse-drawn coach' (ibid., 18). The image of the passive body stuck in traffic or static train is also a familiar dystopian representation in popular culture and yet the sensual experience of commuting is more complex and varied than this implies. Much sensation depends upon the qualities and dispositions of individual commuting bodies and the affordances of vehicles and space, including the qualities of the engine, the smoothness of the road or railway, and the condition of seating. However, as previously mentioned, this body need not be conceived as negatively passive and detached, but as comfortable, lulled into a state of kinaesthetic and tactile relaxation by the gentle rhythms of mobility, as the vehicle or train hums along, the telegraph poles swish by and the engine vibrates with a low murmur. In the case of the hybrid car-driver (Urry 2006), sensation is produced through a synthesis of automotive and human rhythms, the interplay between the rhythms of the machine and the responses and bodily rhythms of the driver:

> during the week if it's a sunny day and the windows are down and there's a cool song on the radio/CD, the road is clear ... you feel alive with the corners that you take ... probably a little faster then normal... its just you and the road, anything could happen, you enjoy the ride as much as possible, drinking in the feeling it gives you.

The rhythms of mobility switch between external engagement, intense concentration on progress and reverie. These modes might be further supplemented by pleasing kinaesthetic manoeuvres as drivers swerve around bends at high speed or trains veer sideways. Of course other factors might disrupt regular rhythmic pleasures: hunger obliterates other sensations; after lengthy spells seated bodies suffer cramp and discomfort; the journey ends and we are hurled into the initially radically different non-mobile environment dazed and disoriented or happy to move more dynamically (for example see Robertson (2007, 85) where she leaves her driving position as 'insulated observer' to change a tyre on the Westway in London). Arrythmic experiences of all kinds disrupt rhythmic calm – tyres burst, loud passenger announcements disrupt daydreams and others drive dangerously.

Some accounts overemphasise the visual dimensions of mobile sensibilities, regarding vision as disembodied, minimal and over-determined by signage, or characterised by the objectification of passing 'landscape'. Yet the mobile gaze can be constituted in numerous ways, perhaps consecutively catching the blur of fast motion, lingering on a roadside icon, travelling empathetically and synaesthetically into the surrounding countryside, glimpsing potential dangers and glancing sight of passing vehicles, anticipating forthcoming signs, and glazing over in reverie. As Degen et al. (2008) have remarked, vision is constituted by multiple modes of looking and Robertson (2007) discusses the movement from the panoramic to closer encounters with space on the elevated highway of the Westway. Accordingly, a gaze activated by habitual passage through familiar space is characterised by a rhythm composed of 'glimpses, transitions, occlusions, and topographies as they unfold around perceiving bodies in motion' (Buscher 2004, 294). It is also inapposite to present a mobile gaze as always purely conditioned by optical disembodiment, for vision is intimately linked to other sensations (Latham 1999), as with the imaginative and empathetic apprehension of space passing (Edensor 2003), but this may extend 'beyond tangible objects to abstractions, virtualities and potentialities' (Shields 2004, 24), making such mobile experience simultaneously regular and open ended.

Commuting, along with other contemporary forms of mechanised mobility, is often understood as linear passage through space, yet this directional flow should not lead to the assumption that this effaces any experience of other coinciding flows and rhythms. As mentioned, all spaces are dynamic and continually pulse with a multitude of co-existing rhythms and flows. Certain cross-cutting rhythms might be particularly evident on a journey, such as the different paces of other mobile machines and bodies, the rhythmic gusts that tug at vehicles and surrounding vegetation, or the flow of a river passing underneath the road or railtrack. And while the insulated mobile body is likely to remain oblivious to the numerous other rhythms that course through place, moments of stoppage – when the train pulls into the station or the car slows to a halt because of road-works or traffic lights – suddenly reveal multiple other coinciding rhythms, and locations become more placial as passengers crowd on platforms or schoolchildren cross the road. At moments of slowing or stoppage, the commuter temporarily inhabits place, perhaps through familiar acquaintance with ticket office personnel, petrol station attendants, and shop assistants, moments and interactions that are woven into the tailored rhythmic event. This changing experience of rhythm and temporality is complemented by the mysteriously subjective nature of temporal experience which is rarely commensurate with clock time:

> The time passes at varying speeds on the journeys to and from work. On one stretch of road between Sandbach and Crewe it can feel like an eternity to reach the end, on other days it's done in a flash, there is no explanation …

Conclusion

To conclude, I wish to make two further points. Firstly, the rhythms of commuting are part of the routines, scripts and habits that make up quotidian life. Accordingly, commuting is akin to the ambiguities inherent in other mundane practices that produce movement and stasis, conform to powerful temporal and spatial regulation but seek certain freedoms, and find a balance between predictability and possibility. As Gardiner argues, the quotidian contains 'transgressive, sensual and incandescent qualities' (2000, 208) and Harrison likewise maintains that 'in the everyday enactment of the world there is always immanent potential for new possibilities of life' (2000, 498). By exploring the rhythms of the everyday mobile practices of commuting, this peculiar balance between tedium and liberation is revealed.

Secondly, this balance between freedom and constraint is unequally distributed and the enforced rhythms and repetitive hazards that typify particular forms of commuting should be acknowledged. For while spaces, timings, materialities and mobilities are orchestrated to provide 'relatively smooth "corridors"' for some (Sheller and Urry 2006, 213), for others, travelling rhythms are far from smooth. While some commuters experience 'connectivity, centrality, and empowerment', for others there is 'disconnection, social exclusion, and inaudibility' (ibid., 210). Consider, for instance, the poor rail commuters that pack trains to and from Bombay, many, unable to purchase a position inside, clinging on to the trains exterior, with many resultant injuries and fatalities. There is little possibility of introducing productive or leisure time into such journeys. Or consider how rising house prices in London have forced many essential workers to live far from the capital, necessitating long daily journeys to work. There are further inequalities between those able to take fast and efficient forms of transport from well-provisioned middle class areas and inhabitants of areas with sparse public transport, between car-drivers who can and cannot afford to pay road tolls, and between commuters distinguished by their gender (see Law 1999), ethnicity and age, for instance.

The practical and epistemological ordering of mobilities has thus 'involved separating out, classifying and ordering travel practices in relation to their efficiency... (establishing) a hierarchy which not only values some travel practices (direct and uninterrupted) and some travellers (fast, orderly, single purpose) over others but also enables their prioritisation in public space' (Bonham 2006, 58). Clearly, some commuters are privileged above others and commuting often causes subordinate rhythms to be reconfigured.

References

Amin, A. and Thrift, N. (2002), *Cities: Reimagining the Urban* (Cambridge: Polity).

Augé, M. (1995), *Non-Places: Introduction to an Anthropology of Supermodernity* (London: Verso).

Bonham, J. (2006), 'Transport: disciplining the body that travels', *The Sociological Review*, 54(s1): 54–74.

Bull, M. (2001), 'Soundscapes of the car: a critical ethnography of automobile habitation', in D. Miller (ed.) *Car Cultures* (Oxford: Berg), 185–202.

Buscher, M. (2004), 'Vision in motion', *Environment and Planning A*, 3, 281–99.

Crang, M. (2001), 'Rhythms of the city: temporalised space and motion', in J. May and N. Thrift (eds) *Timespace: Geographies of Temporality* (London: Routledge).

Cresswell, T. (2006), *On the Move: Mobility in the Modern Western World* (London: Routledge).

Cronin, A. (2006), 'Advertising and the metabolism of the city: urban space, commodity rhythms', *Environment and Planning D: Society and Space*, 24, 615–32.

Degen, M. (2008), *Sensing Cities* (London: Routledge).

Dodge, M. and Kitchin, R. (2007), 'The automatic management and driving and driving spaces', *Geoforum*, 3, 264–75.

Edensor. T. (2003), 'M6: Junction 19–16: defamiliarising the mundane roadscape', *Space and Culture*, 6(2): 151–68.

Edensor, T. (2004), 'Automobility and National Identity: Representation, Geography and Driving Practice', *Theory, Culture and Society*, 21(4–5): 101–20.

Edensor, T. (2008), 'Mundane hauntings: commuting through the phantasmagoric working class spaces of Manchester, England', *Cultural Geographies*, 15, 313–33.

Edensor, T. and Holloway, J. (2008), 'Rhythmanalysing the Coach Tour: The Ring of Kerry, Ireland', *Transactions of the Institute of British Geographers*, 33, 483–502.

Ford, N. and Brown, D. (2006), *Surfing and Social Theory: Experience, Narrative and Experience of the Dream Glide* (London: Routledge).

Frykman, J. and Löfgren, O. (1996), 'Introduction', in J. Frykman and O. Löfgren (eds) *Force of Habit: Exploring Everyday Culture* (Lund: Lund University Press).

Gardiner, M. (2000), *Critiques of Everyday Life* (London: Routledge).

Graham, S. and Thrift, N. (2007), 'Out of order: understanding repair and maintenance', *Theory, Culture and Society*, 24(3): 1–25.

Harrison, P. (2000), 'Making sense: embodiment and the sensibilities of the everyday', *Environment and Planning D: Society and Space*, 18, 497–517.

Jiron, P. (2007), 'Strategies for mobile place confinement/autonomy: the experience of urban daily mobility in Santiago de Chile', paper presented at the Annual Conference of the Royal Geographical Society – Institute of British Geographers, London.

Kaufmann, V. (2002), *Re-thinking Mobility* (Aldershot: Ashgate).

Latham, A. (1999), 'The power of distraction: distraction, tactility and habit in the work of Walter Benjamin' *Environment and Planning D: Society and Space*, 17, 451–73.

Law, R. (1999), 'Beyond 'women and transport': towards new geographies of gender and daily mobility', *Progress in Human Geography*, 23, 567–88.

Lefebvre, H. (2004), *Rhythmanalysis: Space, Time and Everyday Life* (London: Continuum).

Lyons, G., Jain, J. and Holley, D. (2007), 'The use of travel time by rail passengers in Great Britain', *Transportation Research Part A*, 41, 107–20.

Mels, T. (2004), 'Lineages of a geography of rhythms', in T. Mels (ed.) *Reanimating Places: A Geography of Rhythms* (Aldershot: Ashgate).

Merriman, P. (2004), 'Driving places: Marc Augé, non-places and the geographies of England's M1 motorway', *Theory, Culture and Society*, 21(4–5): 145–67.

Robertson, S. (2007), 'Visions of urban mobility: the Westway, London, England', *Cultural Geographies*, 14, 74–91.

Sennett, R. (1994), *Flesh and Stone* (London: Faber).

Sheller, M. and Urry, J. (2006), 'The new mobilities paradigm', *Environment and Planning A*, 38, 207–26.

Shields, R. (2004), 'Visualicity', *Visual Cultures in Britain*, 5(1): 23–36.

Southerton, D. (2006), 'Analysing the temporal organisation of daily life: social constraints, practices and their allocation', *Sociology*, 40, 435–54.

Spinney, J. (2007), 'Cycling the City: movement, meaning and practice', unpublished PhD Thesis, Department of Geography, Royal Holloway, University of London.

Urry, J. (2006), 'Inhabiting the car', *The Sociological Review*, 54(s1): 17–31.

Chapter 13
Tourist: Moving Places, Becoming Tourist, Becoming Ethnographer

Mike Crang

Introduction

In this chapter I want to look at three interwoven mobilisations around travel and tourism. Perhaps the most obvious is the mobilisation of the destination, where I want to suggest that while tourism is often defined as travelling to somewhere – that sense of where is visited is actually rather less firmly placed on the earth's surface than is often assumed. Second, I want to track the mobilisation of becoming a tourist, looking at the construction of tourism as a specific form and practice of mobility, which is perhaps a constrained and less free roving sense of motion than the term mobility often conjures up. And to tell those stories I want in a third register to tell the story of academic mobility – of being a researcher chasing the two previous mobilised topics. To be clear then, the location I am going to discuss is the Greek Ionian Island of Kefalonia, or to locate the destination in not entirely the same space, Captain Corelli's Island. I am going to look at tourists travelling to that island, whichever one it may have been. The chapter is based on field work mostly in 2004, when I was collaborating with a colleague Penny Travlou, some two years after the release of the movie and a good eight years after the success of the novel of *Captain Corelli's Mandolin* (de Bernières 1994).

My plan is to use the register of the ethnographic confessional to illuminate the former two issues – and say something about research on mobile subjects. I shall begin by reflecting on ways of knowing about mobility, or rather mobile ways of knowing – in part to work upon the chiasm of tourism as a practice of travel to other places that often involves generating knowledge in a specific idiom and ethnography as a practice of knowledge that often involves travel (Crick 1992; Galani-Moutafi 2000). It is, I have argued elsewhere, important to see tourism as a knowledgeable activity, if we are to avoid treating tourists as dupes, but one that is not necessarily producing knowledge of an academically respectable kind (Crang 1999). Equally it is a not uninformative conceit to play with the scandalous suggestion that ethnographer and tourist are, if not the same creature then the same species and are part of the same continuum – that *homo academicus* might be uncomfortably closely related to that embarrassing relative *turistas vulgaris*. This essay rejects ideas that the tourist follows knowledge produced by others, codified in guidebooks and their ilk, while the ethnographer produces new knowledge (as though ethnography was not guided in advance in its own way). Some accounts might replace ethnographer with 'explorer'; others might mediate these categories with that of the traveller – one who perhaps follows others' textual instruction but separates themselves from being with other pleasure seeking travellers (Risse 1998). Sometimes the distinction seems to be the velocity of travel, with superficial and brief excursion opposed to slower, more sedentary immersion in place and with slow rather than quick travel producing 'serious' knowledge. Geography has heavy investments, too, in distinguishing fieldwork from recreation, through practices of observation during travel. I recall sitting in front of my undergraduate director of studies, one of Carl Sauer's former post-doctoral students (just to make my claims on disciplinary filiation), with a slide of the great man up on screen, sat on a hill slope, knapsack by him and I am sure I recall a pipe in his mouth, with the quote below 'Locomotion – the slower the better'. But in each of these schema I suggest we can detect a hierarchy of taste and values transmuted into categories of knowing (Bourdieu's 1988). These hierarchies are often organised in terms of levels of reflexivity – where the serious traveller is both more self-aware yet also concerned with others, while the mere tourist is seemingly unreflexive yet focused upon the pleasures of the self. Valid knowledge is deep, reflexive and acquired slowly, whereas declassee knowledge, if it exists, is superficial, unselfaware and unserious – that stages epistemologically the dubious separation of *logos* and *eros* that Wang links to the modernist ontology of tourism (Minca and Oakes 2006). Indeed, the absence and presence of pleasure, or maybe its constitution becomes a crucial issue in categorising practices of mobility.

Perhaps this social categorisation of knowledge and practices can be illustrated through two examples. First is photography. Tourism has been marked as prime territory for photography, and indeed the camera can almost stand as a marker of the tourist on occasions. An obsession with documenting the personal trip, the capturing of clichéd sights, the conversion of sites into sights to be seen, and the sense of the camera as a barrier between local and tourist – all of these are popular epithets about tourist practice (Crang 1997; Crang 1999). Empirically, one can also see 'travellers' then as possessed of more elaborate practices of photography, and more elaborate cameras, working to produce rather different pictures than tourist snapshots (Redfoot 1984). Over again, many ethnographers eschew cameras, partly in favour of the trusty notebook, but partly also to avoid being labelled as a tourist. Or, possibly, because of the impossibility of using a camera unselfconsciously. Personally I chose not to own a camera for several years, finding it quite difficult to take tourist pictures after studying taking them. So in this chapter the practices of photography and the relationship of academic knowledge to practices of picturing will form a framing as it recounts an attempt to conduct a visual ethnography of what is often seen as a visual practice. Second, as an example of slow travel and knowledge claims, is ethnography. Classic ethnographies tend to be written in a fairly declarative tone, with a subdued presence for the ethnographer – if one at all. There is an extensive critical literature around the textual strategies of producing authoritative knowledge. But for now I would highlight the way that personal accounts of the same studies have often been published as separate volumes for more popular markets, shorn of academic constructions. Thus Anna Grimshaw's ethnographic PhD 'Rizong: a monastic community in Ladakh' could become a brilliant tale of personal discovery and tribulation for a more popular market (Grimshaw 1992). The confessional accounts of the blunders and accidents in research may, if not being consigned to a separate volume, form the preface (sometimes literally), or initial chapter to an otherwise conventional ethnography (Crang and Cook 2007, 8). These productions of knowledge cross scandalously from academic to non-academic, so much so that some academics have used nom de plume to prevent the taint of popular writing about their field sites from infringing on their academic credentials (such as 'Joshua Elliot' on Thailand). I would suggest that any border so heavily policed suggests there is a great deal of traffic that has to be denied. I would not for a minute wish to argue that we can or should flatten all the distinctions and that these are the same activities. But the desperate attempts to detach one from the other seem to speak of repressed pleasures and fears. It might be that we need some sympathetic, in every sense, way of producing knowledge about tourism.

How then to respond to this situation where we are travelling to learn about people who are travelling and learning? My response here is to make the further reflexive step of staging how we travel to learn and about travelling and learning. I am then making this account more reflexive, not to deepen it and contrast it with tourist knowledges but rather to render it more comedic and highlight the play of surfaces, rather than suggest some superior profundity. In a sympathetic way of knowing, I am going to suggest thinking through a visual study of tourism as a way of studying a possibly visual practice of tourism. The case in point is the island of Kefalonia that is the setting of a story, *Captain Corelli's Mandolin*, by Louis de Bernières (1994) which he partly researches from the history texts in the library in the island's capital, Argostoli. His story becomes a film that is shot on the island and both promote images of the island for tourists to visit and photograph, whereupon researchers (namely myself) appear and film them and write up stories about the island, occasionally in Argostoli library. To reflect this imbrication of different knowledge practices one might think through an ethnography that is made up of 'many levels of textualisation [and visualisation] set off by experience [and t]o disentangle interpretative [or analytical] procedures at work as one moves across levels is problematic to say the least' (Van Maanen 1988 cited in Wolfinger 2002, 86). It is difficult to separate out a moment of production of knowledge that can suddenly stand apart from the others. Given the confluence of events and redoubling of practices here, writing about writing, picturing about filming, researching in the archives used for the novel's research, it seemed that a confessional account in Van Maanen's (1988) terms was most appropriate, indeed necessary. A confessional emphasises the story of the research and the process of making knowledge. I make this choice with some reluctance, it has to be said, since I have some sympathy with Pierre Bourdieu's (2003) concern that the reflexivity I am undertaking here risks being an example of the 'diary disease' that would seek 'to substitute the facile delights of self-exploration for the methodical confrontation with the gritty realities of the field' (2003, 251). I am very aware that such personalised accounts have as many traps and tropes as others, such as the bildungsroman of eventual scholarly triumph despite mishap (Cook 2001), or the ethnographer as hero, or indeed as bumbling anti-hero, with the demand to confess failings and show vulnerability to gain authority in inverse measure to the textual abjection of the protagonist. I certainly do want to keep in mind Bourdieu's focus upon the relational constitution of knowledge about this field while rejecting precisely the academic politics in that sense of 'grittiness' validating knowledge – that is of course a trope of distinction for 'hard won knowledge' production against comfortable, passive tourism. I cannot really here boast of 'gritty realities' of the field. All the stories I am telling are of a beautiful island, in a stunning setting, in Europe with an industry designed to cater to visitors. Gritty it ain't. Nor can I disentangle my conduct as a researcher from either my 'personal' or academic auto-biographies, from either my sense of doing tourism or doing academic work.

The possible strategies leave us between what I have before called a conceptual Scylla and Charybdis, where on one side is a relativistic immersion into the play of layers of representation, and, on the other, a position that seeks to peel these representations away to somehow get down to a somehow buried reality, lurking beneath the technologies and apparatus of tourism. This latter we might, after Meaghan Morris, call the bad mirror (nasty tourist representation), good mirror (critical social theoretical representation) approach (Crang 2006, 53). This seems an unappetising choice and one where confronting the interplay of representations becomes a necessity. How might we avoid some analysis set in terms of slow versus fast knowledge production, deep versus shallow, with an economy of serious pursuit versus pleasureable diversion? My answer here is to weave both these tendencies through the narrative, into the play of categories and knowledges – a surficial account moving between these loaded images on all sides.

So it seems then that to stage the production of knowledge of and about Captain Corelli's island I have to start with myself as an academic becoming a tourist to conduct a participant observation of tourism to the island. So I want to begin with myself becoming ethnographer and tourist, then look at the mobilisation of the island onto the printed page and celluloid, before looking at practices of consuming the cinematic scene, especially visual ones, before following how the island went missing, and then so too did the film.

Being an Academic, Being a Tourist

'Dr Crang, this is university finance. It's about this holiday you have booked …'

So I was to research tourism on Kefalonia. Myself and Penny Travlou had spoken about it often enough, we had tried to secure funding often enough. Here was somewhere we could see book, movie and tourism and look at tourist photography. Now we had a grant[1] and could proceed. Penny was already 'in the field', by which I mean she had gone 'home', for the first time in ages, to her parents' summer house on the island. I was to follow with a group of tourists from the North East of England where I live. That was when the doubts set in. Sure a partner raised eyebrows about the fieldwork in mid summer in Greece, sure so too do did several colleagues. Not gritty enough. Not serious enough. The doubts became more explicit with the actual planning of the logistics of fieldwork. I had investigated options and the best deal was in fact to simply buy a package holiday as a means of getting a direct flight and accommodation on the island. Doing this taught me two things. First, that only one company would sell a single person a package – couples were the market. Second, the university procurement policy could not cope. I had to submit two alternate quotes (involving flights to Amsterdam, thence to Athens thence Kefalonia or a bus connection and ferry) to prove that this most definitely was not a holiday, but a cheap and expedient way of conducting academic work.

I was then left with thinking about joining the tourists as most definitely not a holidaymaker. The local airport is somewhere with which I have become quite familiar, but only two of my flights from there, though both to Greece, were for holidays. The division of leisure and work travel is etched in the very layout of the building, with the scheduled airlines in a different atrium than the chartered holiday flights (Figure 13.1), and here I was checking-in in the charter rather than

Figure 13.1   Newcastle airport charter flight check-ins

1   Our grateful thanks go to the British Academy for funding this work.

the scheduled flight side, standing in a long queue looking to see if anyone else was a singleton and feeling really rather out of place amid those dressed to start a summer holiday. The in-flight sales magazine headline ('your holiday starts now!') seemed equally perturbing. Eventually we all took off, each of us on the plane nervously anticipating the various things we had all been thinking about and planning for a long time.

I was left here reflecting on how I felt about tourism, and with an urge to start making notes I profoundly hoped would be profound, partly as a way of telling myself that this was indeed work (as I had promised my long-suffering partner left to look after the household). And I have to confess, I felt rather anxious about it all. Not merely the sense that I must produce something worthwhile and maybe even worthy, but about being mistaken for a tourist – or indeed not being mistaken for one. As I sat there I knew part of this was down to my sense that I was not sure about my relationship to tourism in my own life. I have long had the feeling that I make a rather bad tourist in at least three ways. First, I grew up in a family that did not really do tourism. We lived in a tourist area and knew tourist arrivals by their local pejorative of 'grockels'. Second, having studied holidays as an academic, of course, I know in some senses too much to ever unselfconsciously just 'be a tourist'. I am reminded of Claudio Minca's account of being a host to visiting tourist academics when neither party could 'simply' be host or visitor (Minca and Oakes 2006). Third, and finally, my academic training and proclivities seemed occasionally to make me such an obsessively 'good' tourist as to be a bad one. Thus I do read the guidebooks, and the signs and labels on places, and the brochures, and the fliers. I really do get anxious about missing things that I should see or visit. But every time I ask others, they appear to have rather blithely ignored the guides – or at least not been such slaves to them – been less concerned and generally have thus had a rather better time. I am perhaps living proof that you can take Culler's dictum, that tourists are indeed a great army of unsung semioticians (Culler 1981), too far.

In studying Greek tourism, I was encountering myself, but clearly not myself, and revisiting things I knew as a tourist, but in different ways. I felt distinctly estranged from and not at home with the other tourists – which had something to do with the black Moleskine notebook in my pocket, upon whose very materiality I was hanging a set of increasingly anxious assertions about my ethnographic self. On this chartered flight I wondered how much was invested in the material cultures of travel – myself with a stock of chinos and linen jackets that I had deemed the right compromise of work and setting, with the casual and occasionally garish clothing around me. The sort of alienation, and self-alienation created by fieldwork, by turning your daily life into an object of study, was being compounded by assumed (and desired and needed) senses of social differentiation between myself and the other travellers. Indeed, why else would I still feel the need to exorcise and exercise this distance now?

Producing Places: Where is 'Here'?

We were travelling to our destination. In many senses tourism is about producing destinations. Materially it does so by making places to travel too – through travel links, infrastructure and facilities. Skills and knowledges develop around what visitors may want, while the techniques to meet these needs – from sign writing to bus tour itineraries – are also inscribed into the place, typically drawing from wider Mediterranean and Greek experiences. Socially, it mobilises people to host visitors, with the disciplining of local institutions to a specific market and often flows of labour to support the institutions required to host tourism. Thus an island like Kefalonia sees its population of 'locals' surge as the summer season begins and émigrés who had left the island, and indeed migrant workers move in to service the large number of 'outsiders.' To jump ahead, one tour guide to the island (an émigré British woman) explained it more poetically on a coach tour. She recounted how the island becomes quieter and quieter as the season winds down, with drivers like the one on the coach heading off to drive coaches at ski resorts, 'til just a few inhabitants were left, many, she implied, not being local islanders but British immigrants.

For my purposes, here, what is perhaps more telling is the inscription of an economy of desire onto space. Tourism, we may say, is a 'semiological realisation of space' (Hughes 1998) where the physical landscape is turned into a socially produced space through the inscription of meanings; meanings which incite the desire to visit. To put it another way, a destination becomes such by producing a sense of 'hereness' and becoming a place distinguished from others through its possession of some attribute. Increasingly we might argue that the 'hereness' of destinations are not natural features, but rather socially inscribed values and meanings layered onto the landscape. Even the natural is not always secure in offering a sense of self-sufficient presence to a place. For instance Kefalonia is geologically remarkable, with striking fresh water upwellings and the beautiful Merinissi lake in a collapsed cave now open to the sky. Classic tourist features developed here, from boatmen on the underground lake to waterwheels powered by the upwellings but they are not generally reasons given for visiting, rather they are things to do once one has arrived. Beyond natural features many places have become sacralised and given a sense of presence by things not physically present at the destination. As Barbara Kirshenblatt-Gimblett put it, there is a phantom landscape of associations underlying the one we see, where:

> the production of hereness in the absence of actualities depends increasingly on virtualities [...] so that we travel to actual destinations to experience virtual places. This is one of several principles that free tourism to invent an infinitude of new products. (Kirshenblatt-Gimblett 1998, 169, 171)

This not only means more places can become destinations but also that the anchoring of places in their physical actuality becomes rather more tenuous.

Tourist marketing and circulating discourses produce the place and location called 'Captain Corelli's Island'.

The material case in point is film and movie related tourism destinations. There has emerged a niche industry promoting locations used in films, an academic niche discussing the phenomena and indeed a small industry promoting the use of places as locations as a place marketing strategy to local strategic elites. The trade press such as the Manchester Travel News (28 September 2004) listed the top five film holiday destinations, from the UK, as: 1. New Zealand (*Lord of the Rings*); 2. Cephalonia [sic], Greece (*Captain Corelli's Mandolin*); 3. Thailand (*The Beach*); 4. Malta (*Troy*); 5. Kenya (*Out of Africa*). Such lists are performative as much as informative – they solidify a notion of 'film tourists' linked to 'movie' destinations. In response to this sort of industry discourse tourist organisations such as VisitBritain not only promote film related destinations to potential tourists but also offer services to film companies to find destinations – as does the Greek National Tourist Organisation with its guide to previous and potential locations, *Shooting in Greece*.

If the industry is excited about the ability to endow locations with the lustre of tinseltown, then academics such as I can be seduced by the eerie ontological and epistemological symmetry of the processes. Thus perhaps the most influential analysis of tourism in the 1990s was Urry's (1990) *The Tourist Gaze*. This is not the place to argue over the merits of Urry's thesis, but his account of the production of the extra-ordinary as the object for a trained and cultivated form of seeing, adapted from the Foucauldian medical gaze, led to a focus on the production of sights out of sites and the culture of being a sightseer – a word which, when you think about it, is a wonderfully tautological concept. Destinations, it suggested, were rendered into things that could be apprehended through a specific way of seeing and people were trained in that way of seeing. The other great visual technology of the twentieth century has surely been cinema and the rise of the screen. So now we have two technologies which produce specific forms of spectatorship and objects of vision coming together – one founded on a mobilised spectator, the other a mobilised gaze and immobile spectator (Crang 2002). In this scenario, the world becomes that which can, indeed must, be seen and visual consumptions becomes the means of knowing the world.

That tourism is not as simple as that will become clear. But the power of that idea of a visual process remains as a haunting presence for tourists, industry and academics. One way to begin to see this might be the very malleability that enables the invention of destinations. The use of a site in a story or film adds a virtuality which produces a sense of new 'hereness' to a place. But which place? Empirically this can become a dirty contest within the tourist industry as different places pitch their claims. Perhaps the most celebrated case here is 'Braveheart country' in the Scottish Trossachs, marketed on the back of the film about the region that was itself shot in Ireland (Edensor 1997). If we look at the list from 2004 above we see the film *Troy*, about events in Anatolia, benefited Malta – the location of its filming. And some locations are just commonly mistaken, as where many assumed the final scene of *Thelma and Louise* was shot at the Grand Canyon, whereas it was in Utah (Neumann 1999). Instead of a malleable commodity in the hands of a ruthless industry, we find a fluid and fragile set of associations temporarily fixed and held through constant reworking amidst many possible heterogeneous associations.

Kefalonia offers, it seems, a strong case for the association of film and tourism. The book was written on the island, about the island. The movie was shot on the island (after nearly choosing Corfu) about the island. The setting is an island that is pretty self-contained and easily delimited. Even better than the sort of hermeneutic loops described above, the book sold more than 1.5 million copies before the film, and was described in the Guardian newspaper as 'the ideal beach accessory for the discerning holidaymaker' (29 July 2000) – a book to take on holiday that comes to promote a holiday destination. In the press too it has been credited with launching a tourist boom to the island and newspaper travel columns spoke of 'Corelli-mania' leading to the renaming of bars and coffee shops, the printing of glossy guides to 'Captain Corelli's Island' and even the giving away of a *Rough Guide to Cephallonia* in a national newspaper, with an introduction by the director giving the filming locations. The only fly in the ointment of this perfect exemplar might be the pretty dismal reviews of the film. But even here, scathing reviews would point out the island's scenery as the 'best performance'. So let's begin our journey to Captain Corelli's Island.

Anticipating the Scene

Let me build an ethno-fiction here, of planning out the trip and flipping through the brochures. Dominated by blues and whites – as I look to the Greek sections and the Ionian islands, or under 'ideal for couples', romantic destinations and family destinations – the introductory pieces on Kefalonia begin to assume a familiar pattern:

> Castaway Kefalonia – the island of Captain Corelli fame. (Thomson 2005)

> As fans of *Captain Corelli's Mandolin* will undoubtedly know, Kefalonia consists of peaceful bays, tiny hillside villages, sleepy harbours and, also, some wonderful beach resorts. (MyTravel 2005)

> Kefalonia is a haven for beach lovers with its sand and shingle coves, sheltered bays and inlets … still relatively new to mass tourism, although it has become famous due to the success of the book and film *Captain Corelli's Mandolin* … [Sami:] if you want a quick preview see the movie *Captain Corelli's Mandolin* that was filmed here. (Thomas Cook 2005)

> The setting for the romantic story of *Captain Corelli's Mandolin*, this mountainous isle is the largest of the Ionian cluster. Cliffs and caves, picturesque little ports, sleepy villages in herb scented hills, and beautiful beaches – some with watersports, all combine to create the perfect place and space to chill, unwind or enjoy a family holiday. The old Greece with modern comforts. (Airtours 2006)

Any study of brochures has to set such descriptions in the context of the quantitative dominance of pictures and details of pretty standardised accommodation (Dann 1996), where you have a pattern of one page setting the scene for the island then four-to-five pages of accommodation. So the sense of the destination here blurs from Greece to an Island, to a specific resort. In that context Kefalonia though is notable for the scenic and landscape descriptions that are often entirely absent from other destinations. The island truly is the star. And it is the star in the living room before arrival –as Thomas Cook says, it is possible to preview via the movie.

The island of the brochures is helpfully outlined with parasols for major beaches. Greece as a whole plays on the myth of the untouched Edenic beach (Lencek and Bosker 1998) in its publicity. Empty and populated only by the occasional couple the Greek beach of the posters and brochures offers a chance to 'live the myth' of romantic solitude – to adapt the GNTOs 2005 campaign phrase. Kefalonia trades upon one beach, Myrtos, that has become delocalised by its ubiquitous reproduction in images, appearing as 'the beach' in national campaigns where it is unnamed and unspecified and staged to offer the view *of* the beach rather than *from* it. Alternately on the island, Myrtos Beach is profoundly inscribed in place, with signposts for car hire outside the airport using the beach as the symbol of the island, (Figure 13.2a) and more prosaically road signs greeting travelers with the announcement of the impending approach to 'The famous Greek beach' at a mere 25 km or so distance (Figure 13.2b). Indeed the road is set up with a special viewing point from which you may view the (famous) view of the (famous) beach (Figure 13.2c) – safely 5km travel from getting your toes wet, and as a platform now rather safer than simply stopping on a blind bend rounding a mountain spur, though many, more or less, happily strolled across that road each day. And if you hang around that view point, as I did, taking hour long samples, you might expect to see 18 groups of people stop, including two coach parties (but never more than two at a time thanks to the careful scheduling of different companies) for an average of four minutes for independent travellers and a little longer for coaches to allow for disembarking and reembarking (see Figure 13.2d). One could from all this conclude two things. First, there is indeed a visual economy of incitement and satisfaction where tourists reproduce the promotional image for a mediated experience and visual consumption of the scene rather than actually experience the beach. Second, that doing really dull observational tallies is one way to prove you were not having fun but doing academic work.

Figure 13.2a Express car hire, Lassi in Kefalonia: Myrtos c. 40 km

Figure 13.2b Approaching Myrtos c. 25 km

Figure 13.2c Viewing the beach c. 5 km

Figure 13.2d A tour bus party at the view point overlooking Myrtos

And yet being a successful tourist is not so simple as this reading of the signs might imply. As people sat on the viewing platform, they could indeed marvel at the view – and who would not. They would also comment on being there to get 'the view' that they knew they had to have, with a degree of self-awareness that this was 'the picture' they were meant to take. As the guide on one party I travelled with put it – we did not need to worry because the bus would stop in exactly the best location to let you get 'that picture'. An injunction to get the picture, with which they and others were largely happy to comply – although amongst those without an authoritative guide there were sometimes anxious questions asked whether this was indeed the best point from which to do gain 'the picture'. After taking 'the picture' they typically then looked around and moved off, possibly pausing only to walk round the spur to take an equally stunning view north towards Assos. A few would cast glances at the strange chap with a hat keeping a camcorder cool in the baking sun, filming the view, and their part in it. A chap who showed no sign of being with a party, getting back in the car, or even going to take the other view. Someone who was breaking up the ritual being performed.

Hunting Corelli

Travelling to the island is not the same as going to the site of the movie. While the South and West of the island are more closely tied to mass tourism, the settings for the film are distributed across the north and east of the island. It would be entirely possible, if not probable, to visit 'Corelli's island' and not visit any settings in the book or even more so locations used in the film. To find the latter one needs some guidance, be that from locals, the brochure from the DVD, or a guidebook. One also needs a variety of other material supports to enable one to visit sites. One obvious way might be the Captain Corelli bus tour of the island available from the capital Argostoli. A little investigation would, though, reveal that this differed from all the other island tours principally in having a placard saying 'Captain Corelli's Island' on the front of the bus. Essentially then the major enabler of a dedicated movie tourist here would be the hire car.

Penny and I thus hired a car – since though a local she did not have a car nor indeed drive. Nor had I ever hired a car before as a tourist in Greece, and so we both felt strangely out of role. Myself the chauffeur in a foreign land, her accessing different parts of the island. And off we set to track down movie sites and, hopefully, tourists at them. After some quick discoveries it soon became something of a quest for us to reconstruct and identify locations, former participants, memorabilia and tourists across the island. This quest took us over 900km up hill and down dale around the island. It took us on half-metalled roads (that is metalled on one side with the other left rough), up the highest mountain, onto lanes to deserted coves (only some of which turned out to be correct). Along the way we found some of the framing shots of the movie, forgotten threshing floors that had been settings for dances, into derelict villages and nondescript valley sides that looked plausible and were in the right area for battle scenes. The ethnography was becoming road movie.

We became the ur-type of the movie tourist – leaving no stone unturned, no scene unexamined in our pursuit of all things related to Corelli. We were thus utterly unlike most tourists we encountered. Until one day standing at a ruined village we encountered a middle aged man, 'George,' standing arms akimbo staring at one of only three sign boards on the island that depicted the shooting of the movie. We fell into a conversation and, on the off chance, I happened to ask him if he knew about the movie. And boy did he know about it. He was soon regaling me with places visited and scenes he had tracked down with the aid of the extended features on the DVD. I was surprised and asked him about where he was staying – he mentioned the hotel which he found since it advertised that the directorial staff stayed there. I knew it well, and mentioned the signed picture of Penelope Cruz in the lobby. He said that he had found their boats an excellent way to get to the 'fisherman's cove' seen in the movie. I said I had seen them advertising the boats, and contrasted that with the difficulty in finding one of the inland sites. It was now his turn to be interested – how had we found that, could he find it? And so the exchange went on. This was less of me interviewing him than an exchange between two aficionados. The upshot was that we arranged to go that evening to interview him formally and make sure he was aware of our status as researchers – and to my mortification he had assumed I too was doing what he did which was to choose a (war) movie each year and follow it up as his holiday project.

While 'George' proved a limit case in terms of dedicated Corelli tourists, the model of interaction with tourists often followed the line of us becoming expert guides. Or rather while we wanted opinions from tourists, they would trade that for information from us. Equipped as we were with unhealthy levels of knowledge about the movie and the island, we became guides to what we were studying – with those we asked about whether they had visited locations in turn asking us for information about those they had not. Even our presence had the effect of creating a Corelli effect. Standing in that ruined village where we had met 'George', there was very little to see and even with the interpretative materials it took us some four hours to reconstruct the overall lay out of a film set long since dismantled. However, as we stood there looking and noting, passing cars would slow and stop with people getting out to come and see what we were looking at so intently. If we were not there cars would often slow but then not stop. Indeed, the lack of physical remains of the movie meant that many people visited sites without being aware of them. Thus Antisamos beach was partly remodelled, with a cleared area to be the Italian camp, that is now car parking and a club house. Yet when we spoke to tourists they often made no connection despite it having one of the three signboards:

I: So, have you been to Antisamos?
Tourists: Yes, yes. We've just come back from there.

I: Cause – do you know what that was used for? Did you see the notice board there?

Tourist(f):   No.

I: [The Italian camp scenes with the opera group]…That was all filmed down at Antisamos.

Tourist(f):   All I could see was the ice cream sign. That's a shame because I expected to see something like this down there.

Tourist(m):   It's an ideal location down there isn't it – you know there's no buildings at all.

So the question began to form for us as researchers as to whether we had 'lost' the movie as a focus of the project, since if people were not visiting the sites in the movie, or if they did were unaware of their role, then in what ways, if any, were their practices being shaped by the film or book?

Not Being Movie Tourists, Not Being a Film Destination

So in what ways did the movie shape perceptions of a destination? The movie was promoted in every brochure, tour reps had to see the movie as part of their training package, and mention it in their introductory talks, a bar in Skala screened it every Wednesday, the village in which we were based had Captain Corelli's café, there was a range of movie related postcards, and there were copies of pictures and posters of the movie in many places. Not everywhere, all the time, in your face to be sure, but enough to mean that avoiding the movie entirely would be difficult. To sum up the presence and absence of Corelli we might turn to the example of one young man on a sun lounger on Antisamos beach. As I wandered about the setting of the Italian camp in the movie, conscious I was the only person in long trousers, I came upon him and could not but help notice, with deep excitement, that his choice of reading matter was none other than *Captain Corelli's Mandolin*. An excitement somewhat deflated when it turned out he had no idea that the beach was a film location and had not chosen to come to the island due to the movie. He had come to attend a wedding which friends in Britain had organised as part of a growing niche for romantic Kefalonia – that has grown since the love story of Corelli and Pelagia. But once on the island, the connection made him think the novel was a good choice of reading. The story's presence then was marginal but pervasive. It underlay much of the vocabulary of 'romantic' settings and landscape, it helped set up a sense of authentic island out of time (Tzanelli 2003) where a past world had stopped but was still palpable if not accessible.

For the locals, the film represented something of a trap, and one with which they wrestled. Many in the industry were shocked when told of the marketing of the island through Corelli – a controversial rendering of their history. Many, too, worried that branding in this way was counter-productive, with a villa holidays specialist being quick to comment of her friend's Captain Corelli's café that she had warned him: 'My clients would frankly avoid it' as kitsch and being 'too obvious'. Indeed, tourists often singled out the Café for opprobrium with comments such as 'Captain Corelli's café – yeah too obviously a tourist trap' (Figure 13.3). A view shared by de Bernières himself:

Figure 13.3   Captain Corelli's bar, Agia Effimia

A good friend of mine ... who runs a cafe in Fiskardo, likes to tell me that I have ruined his island. He is only half serious, I hope, but it is a thing that worries me none the less. I was very displeased to see that a bar in Aghia Efimia has abandoned its perfectly good Greek name, and renamed itself 'Captain Corelli's', and I dread the idea that sooner or later there might be Captain Corelli Tours or Pelagia Apartments. I would hate it if Cephallonia were to become as bad as Corfu in places, with rashes of vile discotheques, and bad tavernas full of drunken Brits on two-week, swinish binges. (de Bernières 2001,15)

The Captain's cafe was often used as a symbol of 'bad' movie related tourism – by British media, by tourists, by locals and even by de Bernières. It was if anything atypical but served as a marker that the movie tourism market was rejected by, yet unavoidable to, both sides of the industry.

Departures

As I left Kefalonia, I returned to the airport which was overflowing with tourists with whom I queued outside along the pavement under an awning. I bumped into various interviewees, who stared curiously and a little bit obviously, one of them eventually saying 'I thought you were here doing research not as a tourist?' As the warm Mediterranean night closed around us, with the usual melancholy feeling of the ending of a time apart and the impending return to normality, this mobility did seem specifically touristic with its own rhythms and periodicity. The airport will feel very different months later when I am one of the dozen or so travellers catching the Athens shuttle flight having given a talk to the island's Chamber of Commerce on the touristic marketing of Kefalonia. Even writing that here seems to buttress my academic identity – that is a proper academic mobility. But there, in the long queues of people making the best of the last dregs of the holiday, everything seemed very different.

In response I busied myself with the last parts of research, noting and observing, where to my delight I found in the book stands with the light reading for the trip home – or indeed the arrival –not just copies of the book but the edition with the still from the film on the cover. There it is as a memento of the island. Many informants had spoken of using the film to whet their appetites (it had been screened on a broadcast channel just before my research and their holidays) and others, perhaps prompted by the questions, wondered about using it in the long winter months as a reminder so that their holiday and island would return with them. It would become mobilised with an inaccessible and lost time for the island becoming their lost time. And so the novel seemed a perfect souvenir with which to travel home.

And yet typically for the connection of this film and these mobilities, it was not the focal point of concern. It did not organise the travel but was a background frame. Largely ignored by the hundreds of tourists it was a slightly sad and bathetic

Figure 13.4   Departure Lounge, Kefalonia

reminder of the multiple meanings this travel might have and the failure of one to dominate all the others. It also, then, framed a research project, that found a fascinating island but rather lost the movie. A visual ethnography of a movie that had disappeared from sight, leaving only traces and virtualities – of which one trace, for some of these tourists, was two academics popping up all over the place. Feeling, then, concerned about the 'findings' and 'losings' of the research, feeling rather challenged as to whether I was tourist or ethnographer, I contorted myself to take a shot of the novel there in the airport, framing the crowds (Figure 13.4). Setting up the shot seemed to reinscribe the purpose of the research and my identity as a researcher. And as I knelt, what felt like dozens of bored eyes turned to look at this odd and bizarre behaviour, with expressions as if to ask why anyone would want to take such a picture. And at least I felt then I must be a researcher.

References

Bourdieu, P. (1988), *Homo Academicus* (Cambridge: Polity Press).
Bourdieu, P. (2003), 'Participant Objectivation,' *Journal of the Royal Anthropological Institute* NS 9, 281–94.
Clark, S. (2001), *Captain Corelli's Mandolin: The Illustrated Film Companion* (London: Headline Books).

Cook, I. (2001), 'You Want to Be Careful You Don't End Up Like Ian. He's All Over the Place', in Moss, P (ed.).

Crang, M. (1997), 'Picturing Practices: Research through the Tourist Gaze,' *Progress in Human Geography* 21, 359–74.

Crang, M. (1999), 'Knowing, Tourism and Practices of Vision', in Crouch, D. (ed.).

Crang, M. (2002), 'Rethinking the Observer: Film, Mobility and the construction of the subject', in Cresswell and Dixon (eds).

Crang, M. (2006), 'Circulation and Emplacement: the hollowed out performance of tourism', in Minca, C. and Oakes, T. (eds).

Crang, M. and Cook, I. (2007), *Doing Ethnographies* (London: Sage).

Cresswell, T. and Dixon, D. (eds) (2002), *Engaging Film: Geographies of Mobility and Identity* (Lanham: Rowman & Littlefield).

Crick, M. (1992), 'Ali and Me: an essay in street corner anthropology', in Okely and Callaway (eds).

Crouch, D. (ed.)(1999), *Leisure/Tourism Geographies: Practices and Geographical Knowledge* (London: Routledge).

Culler, J. (1981), 'Semiotics of Tourism', *American Journal of Semiotics* 1, 127–40.

Dann, G. (1996), 'The People of Tourist Brochures', in Selwyn (ed.).

de Bernières, L. (2001), 'Introduction', in Clark, S.

de Bernières, L. (1994), *Captain Corelli's Mandolin* (London: Secker and Warburg).

Edensor, T. (1997), 'National identity and the politics of memory: Remembering Bruce and Wallace in symbolic space', *Environment and Planning D: Society & Space* 15(2): 175–94.

Galani-Moutafi, V. (2000), 'The Self and the Other: Traveler, Ethnographer, Tourist', *Annals of Tourism Research* 27(1): 203–24.

Grimshaw, A. (1992), *Servants of the Buddha: Winter in a Himalayan Convent* (London: Open Letters).

Hughes, G. (1998), Tourism and the Semiological Realization of Space', in Ringer (ed.).

Kirshenblatt-Gimblett, B. (1998), *Destination Culture: Tourism, Museums, and Heritage* (Berkeley: University of California Press).

Lencek, L. and Bosker, G. (1998), *The Beach: The History of Paradise on Earth* (London: Secker & Warburg).

Minca, C. and Oakes, T. (2006), 'Introduction: traveling paradoxes', in Minca, C. and Oakes, T. (eds).

Minca, C. and Oakes, T. (eds) (2006), *Travels in Paradox: Remapping Tourism* (Lanham, MD: Rowman & Littlefield).

Moss, P. (ed.) (2001), *Placing Autobiography in Geography* (Syracuse, NY: Syracuse University Press).

Neumann, M. (1999), *On the Rim: Looking for the Grand Canyon* (Minneapolis: Minnesota University Press).

Okely, J. and Callaway, H. (eds) (1992), *Anthropology and Autobiography* (London: Routledge).

Redfoot, D. (1984), 'Touristic Authenticity, Touristic Angst and Modern Reality', *Qualitative Sociology* 7(4): 291–309.

Ringer, G. (ed.) (1998), *Destinations: Cultural Landscapes of Tourism* (London: Routledge).

Risse, M. (1998), 'White Knee Socks versus Photojournalist Vests: Distinguishing between Travelers and Tourists', in Williams (ed.).

Selwyn, T. (ed.) (1996), *The Tourist Image: Myths and Myth Making in Modern Tourism* (Chichester: Wiley).

Tzanelli, R. (2003), '"Casting" the Neohellenic "Other": Tourism, the culture industry and contemporary Orientalism in "Captain Corelli's Mandolin" (2001)', *Journal of Consumer Culture* 3(2): 217–44.

Urry, J. (1990), *The Tourist Gaze: Leisure and Travel in Contemporary Societies* (London: Sage).

Van Maanen, J. (1988), *Tales of the Field: On Writing Ethnography* (Chicago, IL: University of Chicago Press).

Williams, C. (1998), *Travel Culture: Essays on What Makes Us Go* (Westport, CT: Praeger).

Wolfinger, N. (2002), 'On Writing Fieldnotes: collection strategies and background expectancies', *Qualitative Research* 2(1): 85–95.

Chapter 14
Migrant Worker: Migrant Stories

Elizabeth Lee and Geraldine Pratt

[Translated from Tagalog] My name is Liberty[1] and I am here to discuss everything that happened to me and my family before and after I came to Canada. Today is June 8, 2005. I am a mother with four children. Two boys and two girls. I have been separated from them for almost ten years. I left the Philippines in 1990. I went first to Hong Kong to work and try my first luck in working abroad. My eldest son at that time was in the first year of high school and his siblings were in lower grades. With a heavy heart I had to do this. I worked in Hong Kong because I wanted to make a good future for my kids. Anyway, all parents want this for their children … I have no riches or property that they can inherit, right? Only a good education is what I can afford to give them. Not just education but a good one, which I was able to do by sending them to the university.

[Translated from Spanish] Hello. My name is Faviana and I am a janitor. I came to the United States in 1994 from Zacatecas, Mexico. I am single with two children, two daughters, who live with my mother back home. I want to start by saying that I am here [in the US], but I did not want to come here. Do you know of any mother who wants to leave their children? I could not find a job in Mexico and I could not afford to send my children to school. Or buy food for my family. My mother was getting weaker. Everyday we ate so little – just beans – and that is when I knew I had to cross.

We begin our discussion of migrancy with stories of leaving. These recollections embed the solitary economic migrant within their families. They locate them within national contexts where neoliberal policies have led, in Watts' words (1994), to 'the privatization of everything.' Diminishing state support for basic services such as health and education propel even those with middle-class backgrounds and college degrees into migrancy (Parrenas 2001a, 2005; Pratt 2004). These migrants' journeys begin from somewhere; and they are as much about those who stay and the contexts from which they begin as they are about mobility and relocation. The fixity at the heart of migrant mobility is difficult to see from the privileged vantage point of 'receiving' countries such as the US or Canada. We do not see the children or spouses of migrants who clean our homes, hospitals and offices, or who care for our families because they live far away, beyond the borders of our nations. Relocating solitary labour migrants within their families and national contexts, and tracing the intertwined mobilities and immobilities in their lives, are hugely important. This

---

1   This is the name that Liberty chooses to call herself.

allows us to see more clearly how the reproduction of work and family life in countries such as Canada and the US relies on economic disparity between countries in the global north and south, and how low-cost migrant labour is produced by outsourcing the work of reproducing migrant families to contexts such as the Philippines and Mexico, where costs and standards of living are lower (Ehrenreich and Hochschild 2003). It allows us to see how precarious and inequitable are perceived and actual rights to parent one's child in the intimate spaces of a home, and allows us to gauge how the separation of solitary migrants from their families generates new rounds and new generations of social inequality, of mobility and immobility.

We write across two research projects, about two distinct experiences of migrancy located in different national contexts. Elizabeth writes of Mexican courier women who travel regularly between Los Angeles and Mexico, hand-delivering remittances – of money and goods – to their families and friends in their hometowns. Geraldine writes of Filipina women, many trained as professional nurses and teachers in the Philippines, who come to Canada to work as domestic servants through the Live-in Caregiver Program (LCP). The differences between the two cases are profound; the fact and circumstances of migration reflect different state policies, national histories and economic circumstances. The Canadian LCP goes hand-in-glove with the Philippine government's Labour Export Policy (LEP), oddly intimated by the smooth slippage across acronyms. The North American Free Trade Agreement (NAFTA) held out an alluring promise of a better life for Mexicans. However, Mexicans increasingly cross the border because higher-paying jobs have never made their promised appearance (Bacon 2004; Davis 1992; Harvey 2005). For the roughly 21,000 Filipinas registered in the LCP, illegality or undocumented status – though not unknown – is not the most significant problem (Pratt 2004; Stasiulis and Bakan 2003). Unlike their Filipino counterparts, there were about 10.3 million undocumented migrants in the United States in March 2004. Mexicans by far remain the largest group of undocumented migrants at 5.9 million, or about 57 per cent of the March 2004 estimate (Passel 2005).

Across and even because of these differences, we want to explore some commonalities – an eerie repetition of the same – whereby women with distinctive class histories are funneled through migrancy into a narrow range of commodified women's work: childcare, housekeeping and cleaning. They share a common experience of leaving their children behind and mothering long distance (Parrenas 2001b). In both cases, we can see how this separation (and their children's inability to migrate with them) transmits the migration experience across generations, and produces an inexhaustible migration experience within those families. We document the power of migrancy to equalize, that is, to devalue equally. Educated and uneducated women, those who enter through formalized government programs with promises of citizenship, and those who pass into the nation illegally, find themselves in disturbingly similar circumstances.

Clearly, this discussion is at odds with a celebration of migrancy and mobility as boundless freedom and opportunity. The narratives of mobility and travel across borders that we have been told are characterized less by 'freedom' and choice, as by loss, fear, deprivation, and pain. This point has been made repeatedly by feminists over many years (hooks 1990; Kaplan 1987; Massey 1992; Mitchell 1997; Pratt 2004; Visweswaran 1994). It goes beyond an awareness that class differences dictate who is able to travel freely and who travels under severe constraints, to an examination of the consequences of the constrained choices available to migrants. Many migrants are forced into perpetual migrancy, and this not only tears at their hearts when they repeatedly return and leave their children; it also destabilizes the lives of their children, often producing a new generation of migrant workers. Others are trapped in a state of immobility that separates them from their children. In either case, there is no resting place, no happy ending of settlement and integration.

Rather than convey the routinized violence of migration and its aftermath at the border of nation-states, through invocations of the shocking and the terrible, we consider those experiences/times/spaces/scenes in which violence brought forth by migration can hardly be discerned, for instance when migrant mothers are permitted to freely travel between the US and Mexico or the Philippines and Canada, and yet are forced to leave their children behind; when mundane quotidian family problems and events such as late registration at the university, a boyfriend, or mental illness throw a long-sought after family reunification into jeopardy; when migrant mothers register the pain of caring for another's child when she cannot see her own; when a Living Wage, such as the one ordinanced by the city of Los Angeles, still forces mothers to choose between their monthly rent and food for their families; or when mothers explain daily to their children why it is that, even after working two jobs, she still cannot afford to have them live with her. By defamiliarizing the familiar, and tracing the immobilities in mobilities, we hope to illuminate the violence of the mundane, rather than exploit the shocking spectacle (Hartman 1997).

We develop this analysis through the lives of just two women, Liberty and Faviana, not because their lives are illustrative but because they are in no sense unusual. Liberty and her family were among 27 families interviewed between 2004 and 2007 as part of a project on the reunification of Filipino families after the LCP experience.[2] Faviana was one of 30 courier women who Elizabeth interviewed in

---

2   Elizabeth came to know several of the courier women prior to the interviews while working for a labour union from September 2001–2004. Most of the participants in this study were identified with the help and cooperation of organizers at the union. Announcements of the study were made at accompanied work-site visits and informational flyers were also distributed around the union office. Once the initial announcements had been made, I (Elizabeth) was able to collect an unsystematic sample of research participants by using snowball and chain referrals. The interviews were open-ended and in-depth, typically lasting for an hour. The Filipino family reunion project has been done in collaboration with the Philippine Women Centre of B.C. We drew upon the existing networks of the Philippine Women Centre, and the help of a settlement worker, who provided further contacts. The project is the third collaborative research project with the Philippine Women Centre of B.C., each exploring different moments of the migration experience under the LCP: experiences

Los Angeles during the summer of 2005. Rather than reading across a cross section of many peoples' lives, we resist the violence of abstraction that shapes so much of these women's lives, and choose to render Liberty and Faviana as individuals whose lives are structured by regulations and conditions that are shared by many others.

Liberty's and Faviana's Stories

Liberty's decision in 1990 to try her 'first luck in working abroad' led to a decade of extraordinary immobility, regulated by her conditions of employment. She was on temporary work visas, first in Hong Kong and then in Vancouver, which tied her to particular employers and required that she live in their homes. In fact, it was not until 2004, fully 14 years after leaving the Philippines, that Liberty was able to live outside her employers' homes. She stayed ten years with her Hong Kong employers (seven of these were spent in Hong Kong and three in Vancouver) and considers them to have been excellent employers. Still, she describes a very restricted life:

> I stayed seven years with them [in Hong Kong]. My work is only to take care of their children. It was light and not really very hard. I did not want to go out on Sundays – that is why I did not have too many friends in Hong Kong. They were paying me on Sunday that I would stay and work for them. They would ask me to go with them on Sundays to take care of their [2] children and they would pay me 150 Hong Kong dollars for this extra day of work. So, in a period of one month, I would only take a day off once. The minimum wage at that time was 2,800 Hong Kong dollars [a month]. So, I could earn up to another 150 every week. We would go out on Sundays, went out for lunch, watch movies and visit, eat nice food in plush hotels or restaurants. So, it was a nice experience for me. I did not have to do much – just take care of their children. I said that this is okay for me. And they are also nice people.

Liberty migrated to Canada with these employers, but when they returned to Hong Kong in 2000,[3] she stayed in Canada. She made use of one feature of the Live-in

---

as domestic workers, women's settlement experiences after leaving the LCP, and family reunification (see Pratt 2004; Pratt in collaboration with the PWC 2005).

3   It is worth noting the ease and speed with which these employers, who came as business class immigrants with a great deal of capital, revised their migration decision when they found that business opportunities were not as promising in Canada as they had imagined (and the political situation in Hong Kong had stabilized). Women such as Faviana and Liberty, who find that life in the US or Canada is very hard, are nonetheless often trapped in their migration process, especially given their families dependence on remittances. As a measure of this, another woman interviewed for this project recounted that her husband

Caregiver Program through which her Hong Kong employers had brought her to Canada; that is, after working for 24 months as a live-in domestic worker under the LCP, she was able to apply for permanent resident status and sponsor her children: '[my employers] also wanted to bring me back [to Hong Kong]. This time I told them that I am thankful to them for bringing me here in Canada but I am also looking for opportunities for my kids'. Her experiences with the two subsequent Canadian employers were very bad. With the first elderly couple,

> I only stayed there for six months … I didn't get a full day off because during Saturdays before I go, I have to take the dog out. I then gave them breakfast after which, that's the time that I can leave. Then, whenever they want to go somewhere or attend parties, they would contact me and ask me to come home even if it is still my day off because there is no one to let out and walk the dog. That's why I really don't enjoy my off-days.

During the entire six-month period, Liberty was never entrusted with a key to the house, which bound her closely to both her employers and their house. Liberty has traveled the world to work, first going to Hong Kong on her own and then migrating to Canada with her Hong Kong employers. But her daily life has taken place almost exclusively within the cramped spaces of her employers' homes. This radical immobility – the requirement to live in her employers' homes and restricted rights to access – is integral and not incidental to the government programs that enabled and regulated her mobility.

Faviana's final decision to leave her family in Mexico in 1994 was wrought with anxiety about her ideas of what it meant to be a 'good mother.' She states,

> But they [the daughters] were still so young, in middle school. How could I ever leave them, two girls? Two girls. Their father left us even before they were born. But, if I stayed there in Mexico, it would have been impossible to live. What kind of a mother would do that? The only job in Zacatecas that I could find was housecleaning, but I was receiving very little pay and it would cost me more to travel to these homes. Sometimes I would only have one *peso* in my hand. One *peso*! I even tried some construction and building, but every person in the town also wanted that job. There was nothing. I had to come here [to the US] to work.

After a long journey across the border, Faviana was able to reconnect with an old friend, Lucia, from Zacatecas. Lucia was able to find her a part-time waitress job at a local *taqueria* in Los Angeles (making $4 an hour) and was working hard to

---

did not want her to leave the Philippines to work as a domestic worker. But when she sent home her first pay cheque, her family radically reassessed the situation, and said, 'Oh, it's a big amount. Okay. Just stay there and don't come back. Just send the money.' That's what they said to me.'

find Faviana employment alongside her as a union janitor. 'Lucia allowed for me to stay with her in her house. It was a very small apartment and there were seven or eight of us living there – her family and me and some others. I missed my own family so much. Lucia mentioned that I should try working for the union because the pay was good, even if it was hard work. I was willing to do anything.' Two years later, in 1996, Faviana was able to find a non-union job as a night janitor, this time making $5.15 an hour. 'I worked the night shift from 4 pm to 12 am, after my four hour shift at the *taqueria*. The work was hard, but I would only think of my family in Mexico. I had to continue on.' Six months later, there was a vacancy at Lucia's worksite and Faviana was able to gain employment and join the union. Through a rare employer sponsorship later that year, Faviana is now a permanent resident of the US, though she has not yet been able to bring her children to the United States. 'I do have papers [legal documents], but I do not have my family. I could not afford anything for them while I was there, but I am still not sure how much I can afford for them here. But they are able to eat and go to school in Mexico. That is what I wanted.'

Though Liberty did not complete university, she has two years of university education, and her background is a middle-class one. In fact, she was a stay-at-home mother until her husband went overseas to work and stopped sending home a portion of his wages. Her eldest son had placed highly in the entrance exam to a good private school in Manila; it was her desire to send him to this school and her father's encouragement that led her to defy her husband and migrate to Hong Kong. Despite her education and background, Liberty has done nothing but domestic work since leaving the Philippines 17 years ago. This downward social mobility is common among those who migrate through the LCP (Pratt in collaboration with the Philippine Women Centre 2003), the vast majority of whom have a non-university diploma or at least one university degree. In 2003, for instance, 82 per cent of those who applied for permanent resident status after completing the required 24 months in the LCP had a non-university diploma or at least one university degree (Live-in Caregiver Program Fact Sheet 2005).[4] The common experience of working first in Hong Kong or Singapore before applying for the LCP extends the period in domestic service and thus the period of deskilling, making it even harder to regain a foothold in the profession for which one has been trained (Pratt 2004; Pratt in collaboration with PWC 2005). Mobility spawns immobility (or in this case, downward mobility). If Faviana and Liberty began their migrations from very different places and with very different educational backgrounds and skills, their gender and migration to North America have rendered them as equivalent: as non-skilled workers destined for cleaning or domestic work.

---

4   This statistic essentially describes the educational backgrounds of those coming through the LCP because the majority of Filipinas coming through the LCP apply for permanent resident status and over 90 per cent of those coming through the LCP are Filipina.

Liberty and Faviana also share the terrible circumstance of being forced to leave their children in order to care for them. In Liberty's case, both she and her husband traveled abroad (he to the Middle East, she to Hong Kong) and their three children were left in the care of her parents: 'They were with my parents in Manila. They brought them up.' Her Hong Kong employers eventually hired her husband as their driver; Liberty, aged 40, became pregnant with their fourth child.

> He was conceived in Hong Kong but was delivered in the Philippines. My employers had this superstition that I should not have a child when I am under one roof with them and my husband. So, they sent me home, gave me a six-month vacation with pay while my husband stayed in Hong Kong with them. This was the time when my husband started going astray from our marriage and having his own 'extra curricular activities' [with the Filipino housekeeper in the same household]. My child was three months old when I returned to Hong Kong. You see, I returned to the Philippines when I was seven months pregnant. After I had the baby, I left him when he was three months old. I left him with my parents. I went back to Hong Kong in January, and by March my husband was terminated. …When he returned to the Philippines, he did not even bother meeting our children. He did have a look at our youngest when he was only three months old. After that he never bothered and until now we have not heard from him. We don't know of his whereabouts now. So, before I came to Canada, we were already separated.

Liberty herself only saw her newborn one more time in 11 years, for one month when he was three:

> After this, my employers went back to Hong Kong. I was not able to go back anymore. I had already started to save up money because I would be applying for my landed status and start sponsoring them to come to Canada. So, I did not have any more chance to go back, especially because I didn't anymore have an employer who would give me a free plane ticket.

Faviana's education has been limited to middle school.

> From my memories, I was a good student. My teacher said I was very smart and my parents were happy when I came home from school. My goodness, it has been a long time, but I remember, I learned about painting and drawing and I enjoyed reading books. When I was young, my family was poor. Maybe everyone in my town was poor. But, I never imagined that I would have to work all the time, cleaning, and working outside in the fields, for so long. I had two older brothers, who have now passed, but I know they had to stop going to school to help my father in the cornfields. I was the only girl and they protected me from having to quit school to work. When my eldest brother, Jorge, (sighs, 'My God') died from pneumonia, I chose to help my family. I did not have much

strength to help in the fields, but still we needed the money for food and to pay for our land. Each box of corn was worth money and I tried as much as I could do. I was ten or eleven or something like that. I was my daughter's age. It is not a nice memory, and I never want my daughters to have to stop their education. That is why I am here, to *protect* them, so they never have to work in the fields (emphasis mine).

Faviana, like the many other migrant women I was able to interview, laments having to leave the very children they want to 'protect.' She tries her best to visit her daughters, Lola and Lana, a couple of times a year, but Faviana finds that time spent working abroad in the US and remitting money to her family every month is more beneficial than spending time in Mexico. Faviana is not alone in her experience considering the $26 billion in remittances that migrant workers send back to Mexico every year (Pew Hispanic Center Report 2007).

The girls do not have a father, so I try to make them feel secure. I have not seen him in many years and I am glad that he is not around my children. My mother is like both mother and father, so she is strong. I am so nervous when I plan a visit home. I am so excited to see my daughters and my mother. But, when I am there, I always see what they do not have, like what I have in the US, though it is very modest. I feel like I should give them more and that I should work more. And then, I feel panic. I feel like I should be back in the US making more money for them, so they could live better in Mexico. I think working and sending money from the US is the best way I can help my family. I want to be with them in Mexico, but I cannot help them while I am with them. Sometimes Lola understands, or excuse me, Lana understands better, but they have each other. Who knows?

Faviana has thought about sponsoring her children and her mother to the US, but those possibilities and desires are engulfed in more 'worries.'

I think about having my family here in the US with me, but how can I afford that? It will already cost a lot to have them travel here.[5] Three people. Sometimes I can barely afford myself. I have to pay rent, around $700, I buy food, and I have to have money to take the bus to different jobs. And, I also save. I have to save my money for my family. I am not sure if they will come to the US, maybe when they are older. My mother is getting older and my girls are doing well in their

---

5   It is worth mentioning that the US Citizenship and Immigration Services, a division of the Department of Homeland Security, recently announced that it wants to raise the application fee for citizenship from $330 to $595, and the fee for becoming a legal permanent resident from $325 to $905. Some critics have insisted that the pay increases would 'price the American Dream out of reach for qualified immigrants' who were seeking to become citizens (Gamboa 2007).

school. It is not the right time now, but I am not sure if there will ever be a good time. Who knows? It is expensive to live in Los Angeles for me. It is difficult that my family is far away from here, but it may be easier to afford there in Mexico than here. Things in Mexico are bad economically, but I can afford more for them over there than here. My children can have a lot more in Mexico from the money that I am able to send them every month. About $600 every month.

Faviana's uncertainty of knowing 'if there will ever be a good time' underlines the chronic persistence and inexhaustible nature of her migration process. An 'end' to the migration journey is not afforded to all migrants; for some, pain, anxiety and desperation as a result of migration never quite subsides. Faviana is exercising her 'choice' to potentially sponsor her family at a later date, but such a choice is conditioned by her uncertainty about the 'choices' and opportunities that exist for her and her family in the United States and the stark realities of the relative costs of supporting her family in Los Angeles as compared to Mexico. For some migrants, the decision to receive their families may be as dismal as their decision to leave them behind.

As relatively few Mexicans have enjoyed the legal status to permit them to travel 'freely' between the US and Mexico, Faviana has been called 'lucky' by some of her peers.

Many of my friends and my co-workers call me 'lucky' because I have papers [legal documents to travel]. Not many of us have papers. This means I can travel anytime to Mexico. So, I am happy to be a courier for my friends and this way I can also see my family. To be a courier is not a job or a service solely because of the money, but it is a way to be connected with people in Mexico.

In the same breathe, however, Faviana states,

But, sometimes I feel like I am not lucky. Traveling to Mexico and back to the US is not always a happy experience. I think about my family all day and night, they are my life, but even before I arrive to Mexico, I am already thinking about my departure. I travel to visit them, but I am also traveling away from them. Seeing my family and then leaving them is not lucky. In my opinion, I want to be with them all the time.

Though Faviana enjoys her connection with people in Mexico through her job as a courier, her travels as a courier expose her to the pain of perpetual return and departure, and signal her inability to gather her family in one place.

In May 2005, Liberty was able to sponsor her three youngest children to join her in Canada. But as Faviana foresees in relation to her own family, such reunions are not necessarily easy or entirely happy. In Liberty's words:

Every day I am under pressure. I have to work, to think about the food to put on the table everyday so that they would not complain. And they are also adjusting to the time, the weather and the situation here. They don't have friends around … everything is new to them. And then, I am mad all the time because I am used to a clean place and now … my children … their clothes are all over the place. [They share a one-bedroom basement suite.] And this drives me crazy….They are not happy to be with me. That's how I interpret things here with us.

Each child has been placed in their own position of vulnerability through state regulations that have legislated the long separation and by Liberty's fears about successful sponsorship; we expose the vulnerabilities of the two oldest children. The first to be affected was Liberty's oldest son, who lost his rights to sponsorship when he registered late for his university course in the Philippines (given his age, staying in full-time university was necessary to maintain his status as a dependent): 'His immigration was not approved because, at that time, I asked him to quit school because I was financially strapped and could not afford to pay for his schooling. And it was expensive because there were four of us that I was applying to come. I just could not afford it.' For a sum that would seem trivial to middle class Canadians (several thousand dollars), Liberty was forced by her marginal economic circumstances in Vancouver to forfeit her eldest son's rights to family sponsorship in order to sponsor the other three. In 2007, Liberty had not seen this son for ten years, and she misses him intensely.

Liberty's second eldest daughter has been placed in a different type of vulnerability, one that bares an eerie resemblance to Liberty's own life. She had a child in her fourth year of high school: 'But I continued to study [and completed a university degree in hotel and restaurant management] and the baby was not a major obstacle. My husband is an engineer. He passed the exams and he is now a licensed civil engineer. He is now working in the Philippines.' Liberty, her daughter reported, 'was angry [when she heard the news of her pregnancy] and did not talk to me for several months.' Fearing that this would disqualify her daughter's immigration, Liberty did not disclose to Canadian immigration officials the existence of her daughter's four year-old child. Her daughter has left her child in the care of his father. She has returned to the Philippines to marry him and hopes to sponsor her husband and child to join her in Canada. First working in Vancouver as a caregiver for the elderly and as a cleaner, she now works in retail in a shop at the Vancouver International Airport, a poignant and daily reminder of the distance that separates her from her husband and child.

Luck and Responsibility

Beginning from very different places, the 'globalized' urban metropolis of Manila for Liberty and rural Mexico in the case of Faviana, and endowed with vastly different amounts of 'human capital', Liberty and Faviana nonetheless have been rendered as the same lowly valued labour through the course of their migration. As 'third world women' they have been slotted into jobs of cleaning and care in first world homes and offices. In her analysis of the myth of the disposable third world women 'who evolves into a state of worthlessness', Wright (2006, 2, 5) makes the point that, although the myth is excessive to the concrete lives of real women, individual women are made to 'embody the tangible elements of disposability within their being'. Faviana and Liberty do not embody this myth in exactly the same way. With only a middle school education Faviana has moved from poverty in rural Mexico to a secure unionized job in Los Angeles, while Liberty's current employment is insecure private care-giving and house-keeping jobs continues a precipitous downward social mobility that began in the Philippines. Though they do not occupy the role of devalued third world woman equivalently, their employment in low-skilled feminized labour is testimony to the equalizing impacts of their migration.

Both Faviana and Liberty left countries that are under the firm grip of the IMF and World Bank structural adjustment policies, and neither could earn enough at home to adequately support their children and, in particular, to buy them the education that they see as essential to a promising future. They both left their children in their own mother's care for very long periods of time. During the long years of separation, Liberty's situation more closely resembled that of Faviana's friends who cannot return home to visit their children; in Liberty's case this is not because of her undocumented status but because poverty and the expense of saving for family reunification locked her in place. Differently mobile and immobile, Faviana and Liberty are now situated on two sides of a decision: whether to bring their children to join them in the United States or Canada, or to reap the 'benefits' of uneven economic development – this is to raise their children's economic standing in their home communities in Mexico or the Philippines through remittances. This is a cruel choice, and Liberty's challenging family reunion indicates that the decision is by no means a straightforward one. Reunion placed her children in new and different types of vulnerability. As an indication of the profound disordering of family relations wrought by the migration experience, Liberty has in the past dreamt of possibly reuniting with her Hong Kong employers: 'I told them,' she says, 'to give me more time to bring my children to come to Canada and then, maybe, I would then have time for myself and would return to work for them.' Both cases highlight the ways in which liberal acts, policies and gestures such as family sponsorship and resident 'alien' status can intensify the exercize of state power upon migrants, rather than simply relieve their subordinate condition.

For both women, though they began their migration journeys well over a decade ago (in Liberty's case, almost two decades ago), one could hardly say that they have settled. Faviana traces and retraces her journey between Los Angeles and Zacatecas several times a year; Liberty still yearns to be reunited with her eldest son and her daughter still remains separated from her child. Neither Faviana nor Liberty is a solitary migrant and both their mobility and immobility, and their difficulties and successes in settling are only understandable in the context of their families and migrant communities.

Both Faviana and Liberty use the language of luck to assess their situation. Liberty tried 'her first luck' in migrating to Hong Kong; Faviana weighs just how lucky she is to be able to travel freely across the Mexican/US border, to repeatedly move close to and then away from her family. We have tried to come close to the particularity of Faviana's and Liberty's lives to convey the ways they negotiate their difficult circumstances. We agree with Faviana and Liberty that in many ways the language of luck seems much more apt than neo-liberal discourses of choice and responsibility. We ask that the reader consider their blind luck if they were born in a first world country, the ways that they benefit from the circumstances such as Faviana and Liberty, and the choices and responsibilities that flow from this.

References

Bacon, D. (2004), *The Children of NAFTA: Labor Wars on the U.S./Mexico Border* (Berkeley: University of California Press).

Davis, M. (1992), *City of Quartz: Excavating the Future in Los Angeles* (London: Vintage Publishers).

Ehrenreich, B. and Hochschild, A.R. (2003), *Global Woman: Nannies, Maids, and Sex Workers in the New Economy* (New York: Metropolitan Books).

Gamboa, S. (2007), 'Increases planned in US citizenship, immigration fees: Agency expects to raise $2 billion over two years', *Boston Globe*, 1 February.

Hartman, S. (1997), *Scenes of Subjection: Terror, Slavery, and Self-Making in Nineteenth Century America* (New York: Oxford University Press).

Harvey, D. (2005), *A Brief History of Neoliberalism* (New York: Oxford University Press).

hooks, b. (1990), *Yearning: Race, Gender and Cultural Politics* (Toronto: Between the Lines).

Kaplan, C. (1987) 'Deterritorialization: the rewriting of home and exile in Western feminist discourse' *Cultural Critique*, 6, 187–98.

Live-in Caregiver Program Fact Sheet (2005), Distributed at Government of Canada's Invitational National Roundtable on the Review of the Live-in Caregiver Program, January 13–14, n.a., n.p.

Massey, D. (1992), 'A place called home' *New Formations*, 3, 3–15.

Mitchell, K. (1997), 'Different diasporas and the hype of hybridity' *Environment and Planning D: Society and Space*, 15, 533–53.

Parreñas, R.S. (2001a), *Servants of Globalization: Women, Migration and Domestic Work* (Palo Alto: Stanford University Press).

Parreñas, R.S. (2001b), 'Mothering from a Distance: Emotions, Gender, and Intergenerational Relations in Filipino Transnational Families', *Feminist Studies* 27, 2, 361–90.

Parreñas, R.S. (2005), 'Long distance intimacy: class, gender and intergenerational relations between mothers and children in Filipino transnational families', *Global Networks* 4, 317–36.

Passel, J. (2005), 'Estimates of the Size and Characteristics of the Undocumented Population', in Pew Hispanic Center Report, May 21, 2005.

Pew Hispanic Center Fact Sheet (2007), 'Indicators of Recent Migration Flows from Mexico', May 30, 2007.

Pratt, G. (2004), *Working Feminism* (Edinburgh: University of Edinburgh Press, Philadelphia: Temple University Press).

Pratt, G., in collaboration with the Philippine Women Centre (2005), 'From Migrant to Immigrant: Domestic Workers Settle in Vancouver, Canada' in L. Nelson and J. Seager (eds) *Companion to Feminist Geography* (Oxford: Blackwell), 123–37.

Stasiulis, D. and Bakan, A. (2003), *Negotiating Citizenship: Migrant Women in Canada and the Global System* (Palgrave: University of Toronto)

Visweswaran, K. (1994), *Fictions of Feminist Ethnography* (Minneapolis: University of Minnesota Press).

Watts, M. (1994), 'Development II: The Privatization of Everything' *Progress in Human Geography*, 18, 3, 371–84.

Wright, M. (2006), *Disposable Women and Other Myths of Global Capitalism.* (New York and London: Routledge).

Chapter 15
The Vagrant/Vagabond:
The Curious Career of a Mobile Subject

Tim Cresswell

It is surely a strange fact that one of the central figures in the production of modern mobility is the medieval vagrant. The shadowy figure of the vagrant (or vagabond) stands behind contemporary legal and governmental approaches to moving bodies. He floats among the pages of contemporary social and cultural theorists and haunts the origins of the modern novel. Here I want to trace some of the remarkable effects of the vagrant and to reflect on the centrality of the vagrant's mobility to the production of a number of different kinds of knowledge. Such an account is necessarily episodic as the vagrant appears and disappears. For long stretches of time he (the vagrant is almost always 'he' in the sources I am describing) is invisible, inhabiting the spaces where those of us in polite society fear to look or travel. At other times he is caught in the steely beam of surveillance. Occasionally he simply pops up to thumb his nose at the 'truth already established (Bakhtin 1984, 45). My purpose is to illuminate the paradoxical centrality of the vagabond's existence to understandings of mobility in the modern western world. It follows the intuition expressed by Stallybrass and White (Stallybrass and White 1986) that what is socially marginal is often symbolically central.

While there have always been poor people on the road, the vagrant/vagabond did not become clearly defined (insofar as definitions of such shadowy figures are ever 'clear') until the fifteenth century in Europe. The basic facets of this definition have remained constant for over 500 years. The vagrant is a person who has no established home and drifts from place to place without visible or lawful means of support. At least that is how vagrants have been legally defined. Key here are a lack of place to call home, constant but seemingly aimless motion, and poverty. But to start with a definition is to jump the gun slightly. This process of definition is actually the key. In order to trace this process we need to go back to late medieval Europe.

Vagrant One – The (Il)Legal Vagrant

In a fine tuned evocation of the politics of mobile identities the Bern Council of 1481 decided to expel all poor people who were not citizens of Bern (excluding pilgrims who were asked to move on more politely). A later edict of 1483 reiterated this wish to be freed of the wandering poor, this time picking out those wandering beggars who spoke French for particular loathing. This was repeated in 1503, 1510 and 1515. As well as wayfaring paupers, gypsies, pilgrims and an assortment of other 'travellers' were asked to leave – and to never return (Groebner 2007) Clearly the Council of Bern were upset for over 30 years. They were upset by a group of people who were both poor and mobile. There had, of course, always been poor people in Bern and elsewhere. The problem was that these wayfaring vagrants were not locatable. They produced anxiety because they were not legible within the clear hierarchies and geographies of Medieval Europe. This was a world in which everyone had their place both geographically and socially. The poor were tied to the soil both through back-breaking labour and through law. They belonged to places and it was the responsibility of those places to look after them. The Council of Bern, then, was responding to the mobility of the wandering poor. These were people with out place (sans lieu) and 'masterless men' (Beier 1985).

All of the edicts from 1483 to 1515 in Bern demanded the expulsion of the wandering poor. But as the number of these wanderers increased it became increasingly difficult to tell who they were. They also had to be differentiated from (sometimes) legitimate wanderers such as pilgrims. In a small town, where everyone is known, this is not difficult, but as more and more wanderers turn up this becomes complicated. By 1527 the Council of Bern required all the deserving poor (that is the poor who belonged to Bern) to wear badges identifying them as worthy of alms. The authorities kept lists of all those entitled to wear these badges. By the middle of the sixteenth century similar ordinances could be found across Europe. From 1530, for instance, all those practicing beggars in England were supposed to carry a byllet (ticket) when begging for alms (Aydelotte 1913; Groebner 2007).

Across medieval Europe authorities responded to the presence of the wandering poor by differentiating the worthy (because local) poor from their wayfaring (and thus unworthy) counterparts by issuing forms of identification to those who were legitimate and refusing such identification to those who were not. Needless to say, such a strategy also required forms of regulation in the form of people who could check identification and then punish or expel those who were too mobile. Punishment could be harsh, ranging from expulsion to branding and whipping. In 1571 those who were unfortunate enough to be caught in Bern having already been expelled were branded on the forehead with an iron cross.

In these early appearances of the vagrant we can trace how this mobile subject was brought into being as a nightmare figure for a settled society – 'the advanced troops or guerilla units of post-traditional chaos' as Zygmunt Baumnan has called them (Bauman 1995, 94) . We can see how the vagrant became central to the construction of particular laws. These vagrancy laws would travel across the western world over several hundred years. We can see how particular kinds of mobile subject were given identities through papers, badges and other forms of identification that became necessary to the process of labeling. We can see how practices and technologies of surveillance were brought into being to keep the new enemy in view. We can see how the process of identification took the form of primitive biometrics via branding. All of these are reflected in aspects of life today that continue to identify and bring into being suspicious, mobile subjects – dangerous travelers – alternative mobilities. The vagrant is there in the increasing number of anti-homelessness laws that are proliferating in the twenty-first century (Mitchell 1997). He is there in the biometric passport, he is there too in the iris scanner and fingerprint reader. The vagrant continues to haunt the nightmares of the modern state. Nightmares which frequently feature the threatening traveler and the unworthy wanderer practicing unwanted ingress.

About 400 years after the edicts of Bern two black men, Jimmy Lee Smith and Milton Henry, were waiting, one cold, weekday morning, for a car a friend had promised to lend them on a street in downtown Jacksonville in Florida. They needed the car to apply for much needed employment in a produce company. Smith worked off and on in the produce industry and also helped to organize a local black political group. Henry was an 18-year-old high school student. On that particular morning Smith had no jacket, so they went briefly into a dry cleaning shop in the hope of staying warm. They were soon asked to leave. Still fighting the cold they walked up and down the street looking for their friend. Seeing Smith and Henry pass by their store a number of times the store-owners became wary of the two companions and called the police. Two police officers searched the men and found neither had a weapon. Nevertheless, they were arrested due to the lack of identification on the two men and distrust concerning their story. They were arrested on a charge of vagrancy according to the Jacksonville Ordinance Code 26–57 which read:

> Rogues and vagabonds, or dissolute persons who go about begging, common gamblers, persons who use juggling or unlawful games or plays, common drunkards, common night walkers, thieves, pilferers or pickpockets, traders in stolen property, lewd, wanton and lascivious persons, keepers of gambling places, common railers and brawlers, [405 US 156, 157] persons wandering or strolling around from place to place without any lawful purpose or object, habitual loafers, disorderly persons, persons neglecting all lawful business and habitually spending their time by frequenting houses of ill fame, gaming houses, or places where alcoholic beverages are sold or served, persons able to work but habitually living upon the earnings of their wives or minor children shall be deemed vagrants and, upon conviction in the Municipal Court shall be punished as provided for Class D offenses. (Jacksonville Ordinance Code 26–57 cited in Papachristou v.City of Jacksonville 405 US (1972) (156–7))

The case of Smith and Harris came before the Supreme Court of the Unites States of America on 8 December 1971. They were named in Papachristou v. City of Jacksonville (405 US 156 (1972)) as two of eight defendants who had been convicted in a Florida Municipal Court of violating the vagrancy ordinance. The conviction had been affirmed on appeal by a Florida Circuit Court. Their co-defendants had been convicted on charges of vagrancy as well as 'being a common thief', 'loitering' and 'prowling by auto'.[1] The US Supreme Court overthrew the convictions on the grounds that the ordinance was too vague and encouraged arbitrary arrests at the hands of an 'unfettered' police force. Their decision was informed by a working knowledge of the history of vagrancy law that had been imported wholesale from medieval England.

> The history is an oftentold tale. The breakup of feudal estates in England led to labor shortages which in turn resulted in the Statutes of Laborers, designed to stabilize the labor force by prohibiting increases in wages and prohibiting the movement of workers from their home areas in search of improved conditions. Later vagrancy laws became criminal aspects of the poor laws. The series of laws passed in England on the subject became increasingly severe. ...The conditions which spawned these laws may be gone, but the archaic classifications remain. (Papachristou v. City of Jacksonville, 405 US 156 (1972) (161–2))

In fact vagrancy laws had been used in the United States for over 150 years. Papachristou v. City of Jacksonville has become a well known case because it pointed out the absurdity and archaic nature of such laws that effectively allowed police to decide what kinds of activities would fit the term 'vagrant'. Vagrancy was a crime of identity rather the identification of a particular action. In the years following 1972 most States and Canada abolished vagrancy laws and replaced them with new codes specifying particular kinds of behaviour most often associated with the homeless. The Safe Streets Acts of Ontario and British Columbia passed in 1999, for instance, criminalized 'aggressive' soliciting, and the unsafe disposal of needles and condoms. In effect they reintroduced vagrancy laws through the back door.

Returning to Papachristou v. City of Jacksonville, Judge Douglas, who wrote the decision, referred to an array of historical precedents ranging from the laws of Elizabethan England to the history of American literature. 'The qualification "without any lawful purpose or object" may be a trap for innocent acts. Persons "neglecting all lawful business and habitually spending their time by frequenting ... places where alcoholic beverages are sold or served" would literally embrace many members of golf clubs and city clubs' he wrote (Papachristou v. City of Jacksonville, 405 US 156 (1972) (165)).

The decision in the Supreme Court in Papachristou v. City of Jacksonville, is informed by at least a passing understanding of the role of the vagabond in American literature. The decision affirms the connections between the 'freedom' of the vagabond life and what the judge refers to as the 'amenities' of life in America:

---

1   'Prowling by auto' is not listed in Jacksonville's vagrancy statute but the Florida District Court of Appeal had construed this as variant of 'wandering of strolling from place to place' in a previous case.

The difficulty is that these activities are historically part of the amenities of life as we have known them. They are not mentioned in the Constitution or in the Bill of Rights. These unwritten amenities have been in part responsible for giving our people the feeling of independence and self-confidence, the feeling of creativity. These amenities have dignified the right of dissent and have honored the right to be nonconformists and the right to defy submissiveness. They have encouraged lives of high spirits rather than hushed, suffocating silence.

They are embedded in Walt Whitman's writings, especially in his 'Song of the Open Road.' They are reflected, too, in the spirit of Vachel Lindsay's 'I Want to Go Wandering,' and by Henry D. Thoreau. (Papachristou v. City of Jacksonville, 405 US 156 (1972) (165))

Here, the judge notes the role of the vagabond in American literature. But he needn't have confined himself to American writing. In fact, the vagabond plays a central role in the history of the modern novel. This particularly illustrious part of the vagabond's career began shortly after the edicts in Bern, and not very far away.

Vagrant Two – The Picaresque Vagrant

In 1599, the Spanish novelist Mateo Alemán published the novel, widely believed to be the first 'picaresque' novel (Aleman and Kelly 1924). The word picaresque comes from the Spanish term 'pícaro' – referring to rogues and vagabonds. The picaresque novel features the pícaro living outside of and in opposition to conventional society. The pícaro takes to the road and embraces the vagrant life, critiquing, by his very existence, the norms and morals of society. The pícaro has no faith in material possessions and considers freedom and happiness to be the product of lack of roots and attachments (Cruz 1999; Eoff 1953). In *Guzmán de Alfarache*, Guzmán, like many pícaros, starts his life as the child of a wealthy family who effectively spoil him. This is followed by a downturn in fortunes (his father's death) that leads him to take to the road hoping to feed off his family connections. But soon he is reduced to the status of a beggar as he seeks a living in Madrid. Here he learns to enjoy the freedom that comes with having nothing. His life becomes a critique of the hypocrisy of settled society as he employs all the cunning trickery and tactics noted as the 'weapons of the weak' by Michel de Certeau (de Certeau 1984). The novel then follows Guzmán as he moves in and out of high society, occasionally finding worthwhile employment but always returning to life on the road. Guzmán represents the first of many vagabond (anti) heroes who thumb their noses at established society through a life of amoral trickery on the road. The characteristics of the vagrant as defined legally – mobility and poverty – are used to produce a new literary form and a new form of critique. Just as the vagrant plays a central role in the development of modern state-based attempts at control and surveillance of mobility so he inhabits another form of modern representation – the modern novel.

Most commentators recognize the publication of Defoe's *Robinson Crusoe* in 1719 as the birth of the modern novel (Defoe and Richetti 2001). The picaresque novels, however, were the most important precursor to the novel we now recognize. They were considerably more realistic than the stories of chivalry and noble deeds that characterized the centuries preceding 1599. The vagrant then, a figure who only emerged in Europe during the late fifteenth century, became a central figure in the development of a modern form of literature – the novel. Since 1599 the vagrant has traveled through the pages of the novel form – often as a kind of romantic hero who questions the morals and codes of the status quo through restless mobility. He is there in Mark Twain's *Adventures of Huckleberry Finn* (1884) and Jack Kerouac's *On the Road* (1957). He is there in the classic American road movies such as *Bonnie and Clyde* (1967) and *Easy Rider* (1969). He is there in the music of Bob Dylan and Tom Waits. He is there, as the judge nearly 400 years later notes, in Walt Whitman and Henry Thoreau.

While the vagrant quickly established himself as a central character in the new art form of the novel (ironically most often thought of as the literary form par excellence of the new bourgeoisie) he also moved easily into other counter-cultural forms of expression that arose in the twentieth century. Consider popular music and film.

Bob Dylan frequently evokes the vagabond/vagrant figure (sometimes reclassified as tramp or hobo) as a way of critiquing elements of society in 1960s America. Sometimes, in his early topical songs, this is a straightforward use of the vagrants' circumstances to suggest the uncaring and unequal nature of American society. In *Only hobo* (1963) for instance, Dylan tells the story of the body of a hobo lying on a sidewalk, ignored by the people passing by. In later work the tables are turned as the restless vagrant takes a more picaresque role and asks unanswerable questions of those higher up in society. The wandering figure is used to underline the hypocrisy of those above him, sometimes switching roles in order to reveal something of his life to others. Such a reversal is offered at the end of the song, *It's All Over Now, Baby Blue* (1965) for instance, where 'the vagabond is rapping at your door, standing in the clothes that you once wore.' Here the vagabond suggests a change of status and change, or process in general. You cannot be sure that things are stable and will stay as they are. Like the vagabond you have to constantly start anew. The best known instance of this, however, is in *Like a Rolling Stone* (1965), a song which relates a reversal of fortunes for a high society woman. At the outset she is defined in relation to the bums who she throws dimes to as she passes them on the street. For some unspecified reason the central figure in the song is left 'scrounging for [her] next meal' by the end of the first verse. 'How does it feel' the chorus asks 'to be without a home?' The bum that 'Miss Lonely' threw coins to returns in two of the remaining three verses. In the second verse he is the 'mystery tramp'. Here the fortunes are reversed and Miss Lonely, who no one has ever taught 'how to live on the street' finds herself in a position of weakness having to make a deal. By the final verse the vagabond figure returns as 'Napoleon in rags', who Miss Lonely used to laugh at but now cannot refuse as she, like the vagabond, has nothing to lose. Here, at the end of the song the positions are completely reversed and Miss Lonely, and the high society that she represents ('Princess on the steeple and the pretty people… drinkin', thinkin' that they got it made') becomes as invisible as the bums on the street. Dylan's vagabond is radically uncommitted, unwilling to fully invest in values of permanence and stability. He signifies change in time and space.

The vagabond is always where he shouldn't be. He can be found, for instance, dancing wildly around the house of Sir John Soane in London. Sir John Soane was born the son of a bricklayer in 1753, in London. Soane became an architect and designed this house as place for his wildly diverse collection of antiques and works of art which he spent the last years of his life rearranging. On his death he left the house as a museum for amateurs and students. It is full of the spoils of empire, a collection of a kind of wealthy tourist.

The house is the setting for a two screen video installation by artist Isaac Julien called *Vagabondia*. On the two screens we see the house full of the spoils of empire arranged in erratic fashion. We also see a black conservator carefully inhabiting the space, in charge of the arrangement of things. She imagines a number of figures haunting the house.

One of these is the vagabond, also black, who gives his name to the piece. While the conservator moves carefully around the space which is doubled and mirrored by the two screens, the vagabond bounces off of walls and dances in a jagged and irregular way – an embodiment of the post-colonial picaresque. To Nikos Papastergaidis 'The trickster figure in Vagabondia recalls both the trance-like state of colonial cultural adaptations and the nautical experience of swaying melancholy. Mixture seems to come together in a drunken haze and loose kneed swagger' (Papastergaidas 2005, 45). As Elliot and Kennedy put it:

> the Trickster (vagabond) seems both to belong to the uncanny space of Soane's museum, becoming yet another exotic object, and to disrupt its peace and quiet, bringing unpredictable life into the mausoleum. We are uncertain whether he represents a real figure or exists only in the imagination of the Conservator, who watches him dancing in the Monk's Parlour from the safe distance of the Picture Room's raised gallery. (Elliot and Kennedy 2006, n.p.)

Here the vagabond becomes black. His mobility, always irregular and unpredictable, is darkened. The vagabond is still the figure made placeless by the enclosure acts who then goes on to become the character at the centre of a moral panic – an Elizabethan folk devil. But he is also black, a figure of empire, perhaps someone that Soane encountered on his more comfortable travels. He is out of place and out of time and his multiple transgressions encourage us to make connections between the house, the collection and the rest of the world. He is the vagabond who alerts us to the repressed histories and geographies of colonialism. One of the inspirations for the black vagabond in *Vagabondia* is a cartoon print in a series by John Thomas Smith from London in 1817. The collection of cartoons called *Vagabondiana*, represented various well-known vagabonds on the streets of London (Smith 1970 [1817]). Two of these were black and one, who begged in Covent Garden, wore a ship on his head (Figure 15.1).

As Elliot and Kennedy put it, 'the character's ghostly presence conjures up the spectre of colonized subjects and their cultures, many of which Soane collected and objectified through the profits of Empire.' And he does his through his movements.

So while the vagrant is clearly implicated in the dystopian dreams of a modern surveillence state through the writing of laws, the techniques of identification

Figure 15.1   Illustration from John Thomas Smith (1817) *Vagabondiana* ©
Museum of London

and the marking of bodies; he is equally clearly implicated in the more utopian dreams of the various kinds of cultural engagements that emerged within art forms ranging from the novel of 1599 to the popular music of the late twentieth century and conceptual video art. To Julien and Dylan, as much as to Mateo Alemán the vagrant is a figure of inspiration, a role model, an outlaw who asks important questions of established truths. The vagrant then, floats through the centuries as both a nightmare figure of chaos and, simultaneously, as a romantic figure of freedom and critique. In Papachristou v. City of Jacksonville Judge Douglas brings the romantic vagabond in to cancel out the shadow of the nightmare vagabond. How can you persecute this figure so central to American literature (and, indeed, culture) he asks?

Vagrant Three – The Theoretical Vagrant

There is another site that has been traversed by the wanderings of the vagrant. That is the world of social and cultural theory marked. The vagrant/vagabond has followed well-trodden paths in recent accounts of modern and postmodern life.

The vagabond makes occasional appearances, alongside his cousin, the nomad, in the pages of the *Thousand Plateaus* (1987) of Deleuze and Guatarri. They connect the term vagabond to the word 'vague' in a reflection on the work of the phenomenological philosopher Edmund Husserl. The key passage considers materiality. Materiality usually denotes something relatively tangible and unchanging. It is often contrasted with both the ideal, as a non material world of ideas, or the performative as the ever changing but constantly reiterative world of practice. But Husserl located a different kind of materiality that is distinct from the essences that form the object of phenomenological enquiry and different from simple 'thingness'. Delueze and Guatarri take from Husserl the notion of 'vague essences' or 'essences that are vagabond'.

> We have seen that these vague essences are as distinct for formed thing as they are from formal essences. They constitute fuzzy aggregates. They relate to a corporeality (materiality) that is not to be confused either with an intelligible, formal essentiality of a sensible, formed and perceived thinghood. (Deleuze and Guattari 1987, 407)

Elsewhere Deleuze uses the example of the circle as a 'formal, fixed essence' and objects such as plates, wheels and the sun as examples of 'thingness'. A vagabond essence, on the other hand, is neither of these things. Rather it refers to roundness. Roundness, in turn, suggests a process of becoming which has a circle as its endpoint. Deleuze reflects upon Husserl's vague essences in order to get to a notion of 'vagabond materiality' – a materiality that does not have a pure form and is not finished and achieved. It is a materiality of becoming. We do not need to spend any more time considering Deleuze's aims in the definition of vagabond materiality. My concern here is to relate the role our hero, the vagabond, plays. We have seen how the vagabond carries the threat of somewhere else about him – the threat of not here. This is simultaneously the threat of unaccountable mobility and dangerous ingress. It is not a tourist materiality or a commuter materiality or even, as we might expect, a nomad materiality. Here Deleuze employs the geographical imagination of another kind of spatiality to make his theoretical point about form, matter and process. The vagabond is enrolled to bring a mobile vagueness to the proceedings. A sense of not being quite there yet. The vagabond brings a fuzziness to a world some would like to hold as certain, as fixed, as sedentary. This is picked up by Jane Bennett in an essay on edible materiality. 'Edible materiality', she suggests, 'discloses what Deleuze calls a 'vagabond materiality' derived from a consideration of metallurgy'. For Deleuze, she argues, metal represents a kind of matter that is always awaiting form.

> Playing on the notion of metal as a 'conductor' of electricity, he describes how metal 'conducts' materiality through a series of self-transformations – not a sequential movement from one fixed point to another, but a tumbling of continuous variations with fuzzy borders. Metals do not settle forever into one determinate state; alloys bleed into each other. Their efficacy does not depend only on periods of stability: a certain 'incorporeality' is not incompatible with the ability to act or to produce powerful effects. (Bennett 2007, 135)

Here it is quite clear that the vagueness of vagabond life means that the vagabond can play a role which contradicts notions of 'settling', of 'determinate states' and 'periods of stability'.

The vagabond carries his baggage around with him during his wanders through theory. While Deleuze focussed on the unsettled and vague materiality of the vagabond, others enrol him in more pointedly political projects. Cindi Katz, for instance has described a 'vagabond capitalism' in a way which very deliberately plays on the mixture of mobility and poverty that lies at the heart of vagabond existence.

> The phrase vagabond capitalism puts the vagrancy and dereliction where it belongs – on capitalism, that unsettled, dissolute, irresponsible stalker of the world. It also suggests a threat at the heart of capitalism's vagrancy: that an increasingly global capitalist production can shuck many of its particular commitments to place, most centrally those associated with social reproduction, which is almost always less mobile than production. At worst, this disengagement hurls certain people into forms of vagabondage; at best, it leaves people in all parts of the world struggling to secure the material goods and social practices associated with social reproduction. (Katz 2001, 708–9)

Here the vagabond is both cause and outcome. Capitalism itself is a vagabond, the 'irresponsible stalker' of the world. But the vagabond is also the innocent victim of this irresponsible stalking.

Perhaps the vagabond's most persistent role in theory is in the pages of the work of Zygmunt Bauman over the past twenty years. He pops up first in the book *Legislators and Interpreters* (Bauman 1987) in which Bauman provides an account of the role of intellectuals in the modern world, a role that, he argues, has changed from that of a legislator to that of an interpreter. Key to this transformation, Bauman tells us, was the creation of fear and uncertainty in the late medieval world.

> The strongest fear of all was the horror of a new and ever growing uncertainty. This one was anchored in the margins of the familiar and the habitual, but these margins were beginning to press hard at the boundaries of the world of daily life. These margins were populated with beggars, vagabonds, bohemians; through the glasses of popular fear they appeared as lepers, disease carriers, robbers. They were a threat aimed at the very foundations of modern existence, a threat all the more dreadful for the absence of social, customary skills fit to absorb, neutralize or chase it away. (Bauman 1987, 38)

Social order and control in the mid sixteenth century, Bauman tells us, was based on the collective gaze of the small and sealed community – the kind of place where everybody literally knew their place. The emergence of 'masterless men' in significant numbers made this form of control redundant and necessitated new ways of making people accountable. The vagrant, he argues, was thus at the centre of the modern state. Where once the local community had controlled and disciplined the poor, now the state had to step in. Indeed the state, as we now know it, had to be created.

> In England, as in France, the sixteenth and seventeenth centuries were a time of feverish legislative activity. New legal notions were defined, new areas of legitimate state interests and responsibility charted, new punitive and corrective measures invented. Behind all this flurry of activity stood the sinister spectre of the new social danger: rootless and masterless men, 'dangerous classes' as they would later be called, the vivid and ubiquitous symptom of the crisis of power and social order. (Bauman 1987, 43)

In *Legislators and Interpreters*, then, the vagrant/vagabond plays an almost empirical role. Although Bauman is constructing a theoretical account of the rise of particularly new kinds of modern ordering in a more or less Foucauldian manner, the presence of the vagrant/vagabond in that account is a matter of fact – something that happened in sixteenth century Europe and thus necessitated the State, almshouses, poor laws, prisons and other elements in the geography of the modern. Six years later Bauman's vagrant/vagabond becomes more metaphorical – more a figure for the construction of a theory of postmodern ethics.

Towards the end of *Postmodern Ethics* (Bauman 1993), Bauman introduces us to the mobile pairing of the tourist and the vagabond. The vagabond emerges as an alternative to the favoured mobile subject of post-structural thought – the nomad (Deleuze and Guattari 1986; Kaplan 1996; Cresswell 1997). This metaphor, for Bauman, 'does not survive close scrutiny' (240) as nomads follow habitual paths which form a kind of habitual territory invested with relatively stable sets of meanings. Nomads, in other words, may be mobile, but they are very much in and of place while they move and are therefore 'a flawed metaphor for men and women cast in the postmodern condition' (240). The vagabond, on the other hand:

> … does not know how long he will stay where he is now, and more often than not it will be for him to decide when the stay will come to an end. Once on the move again, he sets his destinations as he goes and as he reads the road signs, but even then he cannot be sure whether he will stop, and for how long, at the next station. What he does know is that more likely than not the stopover will be but temporary. What keeps him on the move is disillusionment with the place of last sojourn and the forever smouldering hope that the next place which he has not visited yet, perhaps the place after next, may be free from faults which repulsed him in the places he has already tasted. Pulled forward by hope untested, pushed from behind by hope frustrated … The vagabond is a pilgrim without destination; a nomad without an itinerary. (Bauman 1993, 240)

Here the vagabond is no longer a fact of history but a metaphor – a metaphor for a rootless and meandering postmodern existence – an existence without foundations. And the vagabond has a mobile alter-ego in the form of the tourist. The tourist is also without place, constantly making provisional homes in the places of the other. But the tourist can choose. The tourist travels for pleasure. The tourist is welcome while the vagabond is constantly expelled. Clearly Bauman is writing metaphorically here. The vagabond and the tourist become ideal types – ends of a continuum of mobile postmodern subjecthood that is becoming the mark of both 'postmodernity' and 'globalization'. This is Bauman's 'liquid modernity' (Bauman 2000) where all that is solid has not quite melted into air but become watery, dislocated, overflowing, provisional, transitory and, most of all, mobile.

The mobility is expanded upon still further in 1998 in Bauman's book, *Globalization: The Human Consequences*. Here the tourist and vagabond merit a whole chapter. Bauman claims that now (in the west, in the soon-to-be twenty-first century) it is the 'access to global mobility which has been raised to the top most rank among the stratifying factors' (Bauman 1998, 87) and that this stratification is exemplified by the tourist-vagabond pairing. The vagabond stands in for all forms of forced mobility and all forms of forced immobility. The vagabond is the by-product of new forms of ordering – the waste of the world.

These are the *vagabonds*; dark vagrant moons reflecting the shine of bright tourist suns and following placidly the planets' orbit; the mutants of postmodern evolution, the monster rejects of the brave new species. (Bauman 1998, 92)

Here the vagabond is the refugee, the asylum seeker, the economic migrant. He is all the kinds of mobile subject who move to escape one place or hope for something better in another. And they are necessitated by the all-consuming lives of the tourists who need the vagabonds to service their very different kinds of mobility. The vagabonds are their domestic servants, their cleaners, their nannies, the staff at the airport and in the hotel. The vagabonds are also those suspected of unwanted ingress, of smuggling, of using up welfare services, of stealing homes, of terrorism.

Conclusions

In Bauman's work then, the vagabond's career takes several turns, from fact of history, to subject of ordering, to metaphor of liquid (post) modernity and back to a figure of threat who strangely resembles the nightmares of our Elizabethan forebears. Bauman's theoretical oeuvre continues a long career of the vagabond who has stumbled from the pages of law to the pages of novels, from the compact disc to the world of contemporary art. The vagabond has had the good fortune to play both hero and villain, to be both threat and salvation. As a mobile subject he has, in one form or another been at the centre (and simultaneously at the margin) of the construction of 400 years of law, of the origins of modern surveillance, the birth of the modern novel, and the genesis of what we now call the State. He has also been busy thumbing his nose at all the kinds of ordering that these presuppose in picaresque form from the origins of the modern novel through to the artistic questioning of the colonial past and present. This, I think you will agree, is an astonishing career for a figure who by definition is shunned, excluded and expelled due to his combination of poverty and mobility.

And clearly it is mobility that lies at the centre of the vagrant's career. It was his mobility that necessitated new laws, regulations and forms of surveillance. It was his mobility that proved so powerful as a critique of established moral geographies. And it is his mobility that makes him a powerful metaphor of a theoretical diagnosis of a mobile world marked by globalization, temporality and transience. The career of the vagrant/vagabond is thus underlined by a vivid geographical imagination – a set of knowledges and imaginings about mobility and what mobility might mean in a modern world. The vagabond presents us with an alternative spatiality. Our attention is directed to this spatiality by the laws of Bern which outlined a particular geographical way of being that was not to be tolerated, that was too distanced from sedentary norms. Our attention is also drawn to this spatiality by the versions of the picaresque that ask questions of the supposed morality of ordered ways of being. And this alternative spatiality is enrolled by theorists such as Deleuze and Bauman to undermine taken for granted beliefs about materiality on the one hand and globalization on the other.

What this alternative spatiality does in each instance is help us to diagnose spatiality in general – the spatiality of the normal and the normative. An alternative is not just another way of being and imagining geographically. By its very nature it has to stand in relation to something which is not alternative. Part of its utility is to make investments in the spatiality of life evident, visible, tangible. This, finally, is the trick played by the radical otherness of the vagabond as he flits through the shadows of modernity. We cannot understand such central things as the state, law, and the novel without comprehending the alternative as part of that. And spatial understandings and practice lie at the heart of this process.

References

Alemán, M. and Kelly, J.F. (1924), *The Rogue or The Life of Guzmán de Alfarache* ([S.l.]: Constable).

Aydelotte, F. (1913), *Elizabethan Rogues and Vagabonds* (Oxford: Clarendon Press).

Bakhtin, M. (1984), *Rabelais and his World* (Bloomington: Indiana University Press).

Bauman, Z. (1987), *Legislators and Interpreters* (Oxford: Polity Press).

Bauman, Z. (1993), *Postmodern Ethics* (Oxford: Blackwell).

Bauman, Z. (1995), *Life in Fragments: Essays in Postmodern Morality* (Oxford: Blackwell).

Bauman, Z. (1998), *Globalization: The Human Consequences* (New York: Columbia University Press).

Bauman, Z. (2000), *Liquid Modernity* (Cambridge: Polity Press).

Beier, A.L. (1985), *Masterless Men: The Vagrancy Problem in England 1560–1640* (London: Methuen).

Bennett, J. (2007), 'Edible Matter', *New Left Review* 45: May/June, 133–45.

de Certeau, M. (1984), *The Practice of Everyday Life* (Berkeley, CA: University of California Press).

Cresswell, T. (1997), 'Imagining the Nomad: Mobility and the Postmodern Primitive', in Benko, G. and Strohmayer, U. (eds) *Space and Social Theory* (Oxford: Blackwell).

Cruz, A.J. (1999), *Discourses of Poverty: Social Reform and the Picaresque Novel in Early Modern Spain* (Toronto: University of Toronto Press).

Defoe, D. and Richetti, J.J. (2001), *Robinson Crusoe* (London: Penguin).

Deleuze, G. and Guattari, F. (1986), *Nomadology: The War Machine* (New York: Semiotext(e)).

Deleuze, G. and Guattari, F. (1987), *A Thousand Plateaus: Capitalism and Schizophrenia* (Minneapolis: University of Minnesota Press).

Elliot, B. and Kennedy, J. (2006), 'Haunting the Artist's House: Sir John Soane's Museum and Issac Julien's Vagabondia', *Image and Narrative*. Available online at http://www.imageandnarrative.be/house_text_museum/elliott_kennedy.htm

Eoff, S. (1953), 'The Picareque Psychology of Guzman de Alfarache', *Hispanic Review* 21(2): 107–19.

Groebner, V. (2007), *Who Are You? Identification, Deception, and Surveillance in Early Modern Europe* (Brooklyn, NY: Zone Books).

Kaplan, C. (1996), *Questions of Travel: Postmodern Discourses of Displacement* (Durham, NC: Duke University Press).

Katz, C. (2001), 'Vagabond Capitalism and the Necessity of Social Reproduction', *Antipode* 33(4): 708–27.

Mitchell, D. (1997), 'The Annihilation of Space by Law: The Roots and Implications of Anti-Homeless Laws in the United States', *Antipode* 29(3): 303–35.

Papastergaidas, N. (2005), 'Hybridity and ambivalence: place, identity in flows in contemporary art and culture', *Theory, Culture and Society* 22(3): 39–64.

Stallybrass, P. and White, A. (1986), *The Politics and Poetics of Transgression* (Ithaca, NY: Cornell University Press).

Chapter 16

Refugees – Performing Distinction: Paradoxical Positionings of the Displaced

Alison Mountz

Introduction

At the same time that contemporary existence is characterized by unprecedented global intimacy and mobility, immigrant-receiving states of the global North police borders and exacerbate differences between themselves and 'others' who struggle to land on sovereign territory. States amass information on displaced persons on the move through interconnected databases linked to smart cards and ever-smarter borders (see Amoore 2006). Public discourse about migrants and refugees grows shrill. The attacks on 11 September 2001 enabled the fortification of the edges of social polarization: a stricter policing of the line between those with mobility and those without. The security of the immigrant receiving state is deployed as an explanatory narrative to police those on the move, and the international border plays a pivotal role in this project; not only in its material location at the frontier between here and there, but in its migration into the discourse of daily life where the identity of the alien subject is constructed. The policing of national borders connects with the policing of identities and subject positions within and beyond the state. This connection is made violently for persons displaced and excluded, on the move between states. Citizenship functions not only as a hierarchical framework that orders the legal status of individuals in relation to the nation-state (e.g. Ong 2006), but as a field of practices imbued with contradictions.

Refugees are persons seeking refuge, made visible as mobile bodies in particular ways. They are a group whose mobility is linked to persecution, displacement, and claims for protection. They are sighted, marked, coded, and forced to move in ways that become encoded in law. By virtue of governing the category of refugee, Catherine Dauvergne (2004) suggests that refugees are always othered legally. According to the 1951 UN *Convention Relating to the Status of Refugees* and its 1967 *Protocol*, a refugee is a person displaced from her country of origin, forced into mobility or 'forced migration'. That person seeks protection and to obtain refuge in another state must prove a 'well-founded fear of persecution' if returned home. Policymakers distinguish between two categories of refugees: those located in camps close to regions of origin, and those who travel – usually by employing a human smuggler – to a country with an asylum or refugee claimant program. In seeking refugee status, a person hopes to make a successful claim, but depends on the state for that success.

Current flows of refugees, irregular migrants, and smuggled migrants are categories that are growing more conflated, and a more general criminalized notion of everyone on the move has emerged (as states reduce numbers and retreat from ideals of multiculturalism). The blurring of immigration with the 'war on terrorism' in the interest of national security has spawned a series of extra-territorial enforcement practices.

'Refugee' refers to a heterogeneous set of people, yet is a term that others, discursively, materially, and legally (Dauvergne 2004). For refugees and refugee claimants, subjectivity and mobility are always intertwined and policed through a series of paradoxical positionings. Refugees and those in search of refuge are articulated paradoxically to the state. This chapter identifies some of the ways that refugees and refugee claimants are placed paradoxically within and beyond the nation-state. I argue that those seeking refuge struggle with performances of distinction. Edward Said (1979) illustrated that pronouncements that other reveal more about the powerful than the disempowered. Performances of citizenship as distinction in times of crisis are central to the policing of bodies, an exercise in sovereignty that blurs inside and out, that links discursive and material locations as a way of keeping those constructed as undesirable, poor, and criminal beyond reaching the rights and privileges that accompany membership.

In order to understand how performances of distinction place and displace refugees, the chapter proceeds with a discussion of the shifting role of borders in the governance of mobility by the modern nation-state *vis-à-vis* refugees and refugee claimants. It then addresses paradoxical subjectivities of refugees crafted through performances of distinction. A case study of Chinese refugee claimants in Canada demonstrates that media representations that othered the group were not in fact distinct or the exception, but rather connected with well-rehearsed, racialized tropes of migration.

Mobility, Displacement and Discipline

Since its inception, the nation-state has regulated one of the most fundamental elements of lived reality: mobility. Those who once resided in border regions were required to affiliate with a state, and hence was born the passport as a mechanism of surveillance (Torpey 2000). As Joe Nevins (2002) demonstrates in his history of the US-Mexico border, this sharpening of citizenship as territorial belonging proved central to the establishment of the identity and sovereignty of the modern state. He illustrates how historically the policing of the US-Mexico border was always performed as a tool of state-building. The meaning of the border shifts spatially and conceptually and is called upon to perform many tasks. One function is to link the regulation of mobility to identity and territory: to link who one is to location, and in so doing policing national borders around identities.

As the pace of human migration and communication intensified throughout the twentieth century, questions arose regarding the power of the nation-state. Michel Foucault (1995) argued that, rather than decline, state power historically shifted from violent, public displays of sovereignty intended to demonstrate repressive rule to more subtle, insidious operations of disciplinary power. The modern nation-state works to gather information on a population, a system of surveillance that becomes integrated into the social body so that individuals identified and ultimately regulated each other (Foucault 1991, 1995).

Scholars have studied the effects of disciplinary power on immigrants whose membership in the nation-state is most conditional and regulated due to their legal status (e.g., Pratt 1998). As Michael Kearney (1991) illustrates with examples of Mexican undocumented migrants in the US, border-crossings criminalize and prefigure experiences within sovereign territory where migrants tend to take low-paying jobs in poor conditions, to avoid police in public space, and otherwise feel only partially included. The designation of legal status to im/migrants is one way in which the state assigns identities, sometimes ambiguously:

> The ambiguity of the alien results from policy and policing which inscribe both of these identities – worker and alien – onto her or his person simultaneously. Being neither fish nor fowl and yet both at the same time, the alien is a highly ambiguous person (1991, 62).

Aihwa Ong (2006) connects migrant identities with the state in what she calls 'graduated zones of sovereignty' wherein migrants occupy various legal relationships that dictate degrees of inclusion, roles in the labor market, inscribing identities onto the body by naming, ordering, recording, and defining identities attached to – or detached from – territory. These practices make some feel invisible to the state whilst others feel paradoxically visible and vulnerable.

Sherene Razack (1999) illustrates how policing along Canadian borders echoes the policing of persons of color located within sovereign territory where the border migrates inward. She identifies a binary discourse of 'good' and 'bad' immigrants. The performance of the distinction of the migrant or refugee serves as moment wherein power is transferred through the assignment of identity (Pratt 1998). Migrants and refugees are ordered, the border invoked to position them legally, economically, and socially in 'multicultural' society. Bodies remain linked to borders of the nation-state, enacted through daily acts performed in the project of state building. The border hovers ubiquitously, following one through daily life where discursive acts produce material realities.

For refugees, status is doubly bound to nation-states. They are displaced and ask for protection from a state where they must prove well-founded fear of persecution and lack of protection by the state where they hold or once held citizenship. Some may be excluded by multiple states: displaced from country of origin and then from countries of transit and destination by the policing of the category of refugee. The status of refugee links potential inclusion to previous exclusion, this paradoxical location proving necessary for the membership in the nation-state. Refugee status is also discursively linked to Razack's (1999) 'good' and 'bad' immigrant. The 'good' refugee fits into the definition prescribed by the Convention. The 'bad' refugee will not and is instead positioned as attempting to 'cheat' the system.

States are using these binaries as they increase the percentage of 'good' refugees that they select for resettlement from camps abroad and *decrease* the number of individuals who arrive in sovereign territory of their own accord to make a claim for asylum or refugee status. States do not grant refugee status to everyone, and in their decision-making, they often pit the agency of the state against that of the refugee. Those who arrive of their own volition are called 'spontaneous arrivals', and they are punished for exercising their own agency. In the positioning of refugees as persons in need of assistance, what Ghassan Hage (1998) refers to as the 'gift' of citizenship, asylum-seekers are chastised for exercising agency in transnational mobility.

Over the last twenty years, western countries with managed refugee resettlement programs – the US, Australia, and Canada – have developed sophisticated border enforcement practices. Among these is a practice called 'interdiction' wherein civil servants work informally abroad in human smuggling 'hot spots' like airports to stop potential refugee claimants from reaching sovereign territory (Citizenship and Immigration Canada 2001). This geographical preclusion in the form of border enforcement offshore always constitutes policing of the category and right to access refuge in the form of asylum programs. Asylum seekers thus occupy a paradoxical space: they are, by definition, those who seek protection, but at the same time are punished and less likely to be granted status as a result of having acted to access human and legal rights.

Paradoxical Subjectivities

The assignment of identities to migrants and refugees takes cues from policy, but is enacted and reified in daily discourse. Judith Butler (1997) frames linguistic alienation through the concept of 'speech acts,' by which she means the utterance of categories and the movement of power through discourse. Butler's (1997) conception of linguistic vulnerability operates through the state where cascading binaries are mobilized to exclude. Mobile bodies in search of refuge prove a key site where the nation-state is performed in daily discourse.

In 1999, the federal government of Canada intercepted four boats carrying approximately 600 migrants smuggled from Fujian, China.[1] They were stopped off the coast of British Columbia, believed to be en route to the United States.

---

1 This is part of a larger project that entailed participant observation, interviews, and archival research with Citizenship and Immigration Canada to study the bureaucratic response to human smuggling (Mountz 2010).

Once intercepted, they made refugee claims in Canada, thus becoming refugee claimants. After many released from the first boat abandoned their claims, the federal government detained those intercepted on ensuing ships while their claims were processed. Ultimately, less than five per cent had their claims accepted in Canada, and most were repatriated to China. This proved a particularly shrill moment during which Canadian national identities and values were embedded in and articulated through border enforcement, immigration policy, and the refugee claimant process. The reception and repulsion of the claimants in public discourse demonstrate the performance of distinction, the gestures through which they were scripted as alien to the citizen, other to the state and its legal, economic migrants and 'genuine' refugees.

It is worth noting the contrasting reception and discourse surrounding the arrival of 5,000 Bosnian refugees resettled in Canada only months prior to the arrivals from China. Several factors distinguished representation and reception of these groups. Bosnians were victims of a political conflict with which Canadians were familiar. Like business immigrants whose migrations are facilitated, they were brought to Canada by the federal government, their migration 'managed' in orderly fashion (Hyndman 2000). They were chosen by authorities from the country where they would be resettled, rather than having chosen mobility themselves, acted upon in a humanitarian fashion, rather than choosing to exercise their own mobility. They were also not part of the racialized 'Asian invasion' That would follow and echo racialized representations of earlier migrations (Anderson 1991; Bhandar 2008). The Canadian public and civil servants exuded pride in their facilitation of the humanitarian landing of Bosnian refugees. Soon, they would exhibit pride in the exclusion and repatriation of Chinese migrants.[2]

In contrast, the 1999 boat arrivals, though much smaller in number, grated against the popular geographical imagination of Canada as humanitarian, diplomatic state that exercises order and good governance. The interception of 'boat migrants' relayed the image of illegitimate crossing and abuse of the system: something all together *disorderly* (see Cresswell 1997, 2006). As bureaucrats struggled with the public narrative that they were ineffective (Hier and Greenberg 2002), refugee claimants struggled with the narrative that they were 'bogus'. Through metaphors of disease, natural disaster, and criminality, they were positioned as a threat to national security. This movement was generalized to a broader critique of the federal government, and the story went that unexpected and unmanaged arrivals opened Canada's immigration and refugee programs to abuse. Part and parcel was the portrayal of the Canadian federal government as incapable of controlling its borders, overwhelmed by human migration, and guilty of weak laws and a flawed bureaucracy (Greenberg 2000).

---

2   This pride was evident in interviews (Mountz 2009) and in public discourse when it was leaked that a manager in CIC had distributed collared golf shirts that said 'Class of 90' to employees involved in the repatriation of 90 of the refugee claimants by airplane.

Media representations proved paramount not only to public opinion on migration (Rivers and Associates 2000), but also to the daily orientation and operation of the federal bureaucracy in its response to refugee claimants (Mountz 2010). The process of identification that took place in the media, far from superfluous to the federal position on human smuggling, provided a narrative that explained and justified the response. The claimants' stories were challenged in interviews, formal hearings in the tribunals of the Immigration and Refugee Board, and in legal negotiations. To hail 'bogus' refugees into being through interpellation (Bassel 2008), while at the same time questioning their stories and intent is to place them paradoxically in relation to the state. They were, at once, asking for protection from the state and demonized for exercising agency to 'cut the queue' to enter Canada ahead of an imaginary line of legal immigrants. The discursive battles over this group of refugee claimants corresponded with the institutional geography of their detention. Paradoxical subjectivities corresponded with paradoxical geographies. Though seeking refuge, they were detained remotely in the interior of British Columbia where access to refugee programs was mediated by distance. They found themselves neither here nor there, included through exclusion in Giorgio Agamben's parlance.

Distinction

Benedict Anderson (1983) has articulated the ways that nation-states are communities imagined through public discourse. The media perform state and nation *vis-à-vis* the refugee not only with discussions of border enforcement and refugee policies, but through the language of alterity, through the utterance of who *they* are. These performances of distinction articulate refugees and claimants outside of the nation-state, simultaneously homogenizing the group with gestures of sameness. Talk about refugees, in turn, performs the state, suggesting broadly what it means to be Canadian by demonstrating what it means for a group to be *not* Canadian.

The language of migration has always been one of vulnerability, belonging, and exclusion (Ellis and Wright 1998; Cresswell 2006). Utterances place refugees and claimants outside of the state in order to rationalize projects that benefit those communities imagined on the inside. For the refugee claimants from Fujian, these terms were constituted by linguistic distinctions. When the boats were intercepted, the Canadian media were saturated with images of a group that came to be known as 'the boat migrants.' As with the Vietnamese refugees resettled in the 1970s, the 'boat migrants' identities blurred with their mode of transport, and they never quite left behind the hyper-visual moment of arrival. Through the act of naming, they were distinguished by mode of travel. The media also emphasized the migrant body, focusing early coverage on the potential for disease, including tuberculosis and hepatitis, and other medical conditions like malnutrition, dehydration, and hypothermia. Migrants were portrayed as carriers of disease, exaggerated in the press by images of people crowded on boats in close proximity and unsanitary conditions. One notable example in a daily local paper, *The Province*, carried an image of migrants crowded on deck with the large headline 'Quarantined' (*The Province*, 21 July 1999). Hence emerged racialized tropes of disease and 'yellow peril' that surrounded earlier Asian migrations to North America (Anderson 1983; Li 1998).

Media coverage overlooked the reality that this was a relatively small number compared to those smuggled through Canadian airports on a daily basis (see Clarkson 2000). Instead, migrants were described with hyperbolic metaphors of natural disaster, as in the 'flooding' of the nation-state, an oft-repeated trope in immigration stories (Ellis and Wright 1998), encapsulated in the headline 'ENOUGH ALREADY' (*The Province*, 1 September 1999). The numbers were decontextualized and portrayed as crowding. Combined with the direct route from China to Canada, the portrayal of bodies in close proximity to the border tapped into fears of invasion. Hage addresses numerical decontexualization by the media with the use of vague numerical categories like 'too many' as 'categories of spatial management' (1998, 138). He illustrates that such phrases are always racialized, the center of the nation-state imagined as white, with the power to manage 'ethnic' others under the rubric of multiculturalism.

The migrants were also distinguished through the deployment of binaries. Binaries of inside and an outside are constructed to justify the center. One binary emphasized the criminality of entry by way of organized criminal activity, dwelling once again on moments of entry. Constructions of criminality - also a common trope in media coverage of recent immigrant communities - gained momentum throughout the claimants' time in Canada, from their portrayal through security fences under surveillance at detention sites, to their movement in and out of correctional facilities and courts during the next 18 months (McGuinness 2001, 20-21). Mahtani and Mountz (2002) found that 'articles that associated immigration with criminal activity rose significantly in scale and pitch' from 1995 to 2000, peaking with the boat arrivals in 1999. The discourse of criminality ('illegal aliens') accompanied media representations of the boat arrivals saturated with headlines such as 'Detained aliens investigated' (*Vancouver Sun*, 22 July 1999).[3] Such language pitted a Canadian 'us' against a foreign 'them', as in the headline, 'Beware, illegal immigrants. We Canadians can be pretty ruthless' (*The Province*, 13 August 1999). National narratives took shape: these were non-Canadian values.

Whereas other immigrants are applauded for their economic and cultural contributions to a multicultural society or narrated as needing protection, these claimants were punished for taking advantage of Canada's 'generous' refugee policies. The assertion that Canada is too generous resonates with Ghassan Hage's (1998) argument that western states construct citizenship as a gift rather than a

---

3   This language could be viewed as an Americanization and, therefore, harmonization of borders through discourse.

right, hence the 'generosity' of Canada's refugee program. Hage calls detention 'ethnic caging,' the isolation and containment of those who break the rules. Ien Ang argues similarly that the 'hostess' who bestows the invitation into the nation retains the right to decide who enters, to behave as gatekeeper (1999, 198). In this discourse, citizenship is a thing: something bestowed, granted, or taken away. The 'bogus' refugee is positioned and racialized as the person attempting to take something not deserved.

Other binaries were repeated so often in newspapers as to seem irrefutable: these migrants were 'bogus' not genuine, 'economic' – and therefore opportunistic – and not 'political', not deserving but undeserving, not good but bad, not legal but illegal. Binary distinctions and their reification homogenized the group, declared as neither legitimate economic migrants nor legitimate political refugees. They were often presented to the Canadian public as distinct, as a crisis too large for the Canadian government to handle (Hier and Greenburg 2002). Such performances are paradoxical because the state needs and has historically needed migrants for many reasons – to work, populate, invest, defend – but it also needs migrants to expel in order to narrate its center through performances of distinction. Every category policed along the borders of the nation creates an 'other.'

If refugees serve as the other to the economic migrants desired by the state, 'bogus' refugee claimants serve as the other to the 'genuine' or 'Convention' refugee. For Agamben (1998), such exclusion is central to sovereign power. The result is a paradoxical positioning of refugees in relation to the nation-state and a corresponding and exclusionary geography of exception.

Crisis

Human smuggling from China to Canada was neither historically distinctive nor exceptional, but construed as crisis nonetheless. In interviews, civil servants often argued that Canada has the 'right to choose' its migrants and refugees, an entitlement that underscores the battle of volition between state and migrant and the ability to move. The right of the state to choose contradicts the notion of displacement, which is not a matter of volition, and states' responsibilities as signatories to the 1951 *Convention*.

Yet the boat arrivals played in mainstream media as a crisis that positioned claimants as a threat to security, provoking anxiety in the public. In her content analysis of the coverage of the arrivals in four dailies, Sorcha McGuinness (2001) argues that journalists were 'swept along by waves of public opinion, dammed by institutional constraints and caught in an undertow of bias' (2001, 23). Like McGuinness, Hier and Greenberg (2002) point to content, tone, and language in the 'discursive construction of a crisis' with 'a capacity to recruit and mobilize newsreaders as active participants' (2002, 491) in the narrative of who comprised this group of migrants, and who comprised Canada. The arrivals signaled both quantitative and qualitative shifts in coverage, and much of the crisis in the news unfolded as a 'moral panic' wherein the events of the boat arrivals and the numbers of migrants were decontextualized, took on a life of their own, and hence became a 'phenomenon' (Hier and Greenberg 2002, 503).

Images of immigration often narrate the story of the emasculated state rendered powerless by immigration that is out of control, embodied by migrants who materialize in discourse with metaphors of invasion, flood, and waves (Ellis and Wright 1998). Human smuggling and trafficking serve particularly poignant affronts to the nation-state because they entail an illicit undermining of international borders and laws and the movement of bodies as commodities for consumption in the global sex trade and other service economies. The migrants were represented in the media as a challenge to Canadian sovereignty and its ability to police international borders. This discourse of leakiness with regard to North American borders was on the rise for some time and escalated after 9/11.[4]

Cresswell (1997) traces the discourse of seepage surrounding medical, ecological, and bodily metaphors deployed to put people 'out-of-place.' He argues that metaphors organize actions; they 'serve to link order to place and space not only as a descriptive device, but also as a way of thinking and acting' (1997, 331). Some bodies are more visible because of race, class, gender, and legal status, all of which figures prominently in discourse on immigration and is central to decisions about who 'belongs' to the nation-state (Honig 2001). Cresswell (1997) theorizes society as a human body, and leaks as out of place, in need of being cleaned or removed, and this is precisely what happened with narratives and practices of containment with which the government responded to the migrants. In response to pressure from national and international publics to strengthen 'leaky' borders, the federal government communicated a semblance of control to the public in order to counter media representations of a diseased Canadian social body, leaking out of control (see Cresswell 1997; Hage 1998). As such, the border around the nation-state serves as a symbolic expression of the well-being of the Canadian social body.

Discourses of displacement take particular shape in distinct national contexts. Canada, once a nation-state known for more progressive, humanitarian practices in the granting of refugee status, was now portrayed in the media as soft, with the integrity of its refugee program threatened. The boat arrivals catalyzed a notable shift in Canadian public opinion toward immigration, public discourse surrounding immigration, and the political will and capacity of the government to respond. The media produced these images for consumption by an anxious public, the migrants serving symbolically as an expression of a perceived loss of control of borders. Such public anxiety is itself an exercise in power: for those on the inside to police

---

4   The US viewed the border between itself and Mexico as 'leaking' migrants for decades (Nevins 2002). More recently, Canada became the source of such leaks across the northern border of the US. The attacks on 11 September 2001 provoked discussion of vulnerabilities along the border and both countries poured millions of dollars into the fortification of the Canada-US border with new policing agents and technologies.

those outside (Hage 1998; Razack 1999). Refugee subjectivities are detached from a discourse of 'protection,' and bound up in narratives of nationalism and enforcement.

The arrival and eventual 'removal' of most of the Fujianese migrants, coupled with the announcement of new immigration legislation in April 2002, marked the culmination of a decade of tightened controls over immigration, during which time, remarked Razack, 'The criminal attempting to cross our borders featured as a central figure in the discursive management of these new [federal] initiatives' (1999, 160). With the state represented as leaky body, the construction of boundaries around identity is one strategy of containing leaks (Hyndman 2004). Such postures of containment result in powerful material ramifications for refugee claimants and other displaced people, including detention and deportation. The boundaries around identity construction were policed, resulting in tight narratives about human smuggling from China to Canada and the unusual mass detention of refugee claimants.

Crisis is one mode of being and of enacting citizenship and enforcement (Mountz 2010). When the state enters crisis, the bureaucracy becomes highly responsive through its communications that something must be done right away. Whereas enforcement officials narrate the story of needing to intensify policing to respond to increases in human smuggling, research shows the relationship to be reversed: when states increase enforcement, potential asylum seekers are pushed into the hands of smugglers as it becomes more difficult to reach sovereign territory (Nadig 2002). Yet civil servants were reluctant to concede the connection between enforcement and smuggling or that most refugee claimants must employ the services of smugglers in order to arrive 'spontaneously' in sovereign territory.

Exception

Crisis is necessary for sovereign constructions of distinction and suspensions of the law (Agamben 1998). In the Canadian case, policy decisions are made 'on the fly', as civil servants often characterized them in interviews, and the government invests enormous resources in aggressive detention and deportation practices (Mountz 2004). In Agambennian terms, these beget ambiguous 'zones of exception', such as camps (see Perera 2002), and exception ultimately becomes the rule through the topological blurring of inside and out. Agamben's argument about the paradox of sovereignty illuminates the underlying rationale for performative acts of the state, which migrate into the performative acts of individual identities. The exception, the crisis, exists to reify the importance of the citizenry as center: to protect citizens, maintain the integrity of borders through processes of abjection (Kristeva 1982).

Immigration and refugee policies themselves prescribe varying degrees of national belonging, but exclusion also takes place in discursive practices of categorization. The process of identity construction that took place in Canada is consistent with Foucault's theories of biopower and governmentality (1991), wherein the state manages populations by producing identities discursively through practices of classification and categorization, exercises that entail the material inscription of identities onto the body (Pratt 1998). The media contributed to the regulation and surveillance of refugee bodies in relation to popular interpretations of immigration policy. The narrative of the illicit entrant affirmed the story of the violation of what was perceived as a nation already too generous with its immigration policies (cf. Razack 1997, 173). This discourse contributed to the environment in which migrants experienced the refugee determination process.

The discourse in which Canada was scripted as unique also proved not to be the exception in global context. Razack asserts, 'One of the paramount tasks of border control, and the justification for all new initiatives, became the separation of the legitimate asylum seeker or immigrant from those deemed to be illegitimate' (Razack 1999, 160). In contrast with the discourses of globalization that involve flows of capital and elite business professionals across borders, migrants, and especially poor migrants and spontaneous arrivals, are often characterized as leaks and invasions. This narrative parallels trends in Australia and the European Union where it was also becoming increasingly difficult to be a 'legitimate' refugee. Increasingly, in North America, Europe, and Australia, refugees and asylum seekers are drawn into the narrative about the loss of state sovereignty and viewed as somehow 'economic' and therefore not 'political' or genuine. Constructions of illegality in the media have led to the rise of 'the bogus refugee'. At the same time that states are tightening controls on refugee movements, the media are delegitimizing those individuals that do succeed in making refugee claims.

Western nation-states have shifted their view toward asylum-seekers, scripting them as bogus, and subjecting them to securitization over protection. This crisis emerges not only in the logistical control of mobility through policy, law, or detention, but also through Butler's (1997) speech acts. Performances of distinction include and exclude by naming and locating. The tightening of control corresponds with this *global* sharpening of discourse about refugees and the reduction of their numbers in western states: the rise of the bogus refugee is occurring, not coincidentally, everywhere that the largest per capita western receiving states – Australia, Canada, the United States – simultaneously produce and exclude what they call 'irregular' migrations.

Amid discursive crises at home, states exercise more aggressive enforcement practices offshore that preclude access to asylum and refugee claims processes. The detention industry is growing and becoming more remote, with creative geographies of interception and detention on Guam and Guantánamo, Nauru and Christmas Island. These offshore locales correspond with Agamben's paradoxical spaces of sovereignty. There, claimants occupy paradoxical positions in relation to the nation-state both in public discourse geographically and legally, *who* they are determined by their location: not here, not deserving, not refugees (Mountz 2010). This designation in turn influences access to legal representation and human rights.

In crisis, national security is performed to safeguard against 'the bogus refugee'. Boundaries around nationalism are summoned for particular purposes, and crisis is put to work. As with the conjuring of the border in everyday life, these violent moments of exception become everyday practices of exclusion. State performances of crisis become intense expressions of sovereignty. With citizenship conceived as a field of practice, language functions as daily mode of inclusion and exclusion. Therein, refugees and asylum-seekers are performed outside of the state. Even once inside sovereign territory, they occupy Agamben's paradoxical space, both and yet neither inside nor out. Hage's (1998) gift of citizenship is given and taken away. Binaries are deployed to shift refugees from outsider to alien, from legitimate to criminal, deserving to undeserving, inside to out. Categorization lies with the purview of the nation-state, and crisis provides the context necessary for exception and exclusion.

Agamben suggests that exception becomes the rule. The rise of the bogus refugee in a variety of national contexts is no longer the exception, as in the abuse of the 'generous' Canadian refugee program, but the rule, as in the rise of the bogus refugee in immigrant-receiving states. In the conflation of categories to criminalize anyone on the move, refugees too are being criminalized. The terms 'asylum seeker' and 'refugee claimant' no longer conjures an individual in need of humanitarian aid, but rather an economic migrant attempting to exert her own agency, and in so doing, undermining the power of the state to 'chose.' Those who cross onto sovereign territory are criminal, contrasted with those managed at a distance. Accordingly, trends in refugee management all relate to containment of migrants closer to countries of origin, including geographical strategies of safe third country agreements, remote detention practices, and 'protection in region of origin'. The now shrill climate of border enforcement that acts on behalf of those who comfortably occupy the state enables discourse to become reality; exception to become rule; crisis the norm; distinction exclusion.

Conclusions

Nation-states position migrants and refugees paradoxically in order to bolster the tenuous foundations of nationhood and nationalism, if fleetingly. Refugees and asylum seekers are positioned paradoxically in relation to the state: the former in need of protection, the latter admonished for their efforts to access protection. The blurring of inside and out are part of the crisis, during which time, distinctions masquerade as exceptions before becoming the rule. Through their alienating utterance, distinctions function to exclude by distinguishing between those inside and outside of the sovereign territory of the modern nation-state. Debates have ensued regarding the ways in which the nation-state is being reconfigured and re-spatialized (e.g. Appadurai 2006). As sovereign territory becomes more dispersed, migrant and refugee categories are blurred, and increasingly, everyone more policed by the state.

Disciplinary power operates simultaneously through the policing of borders and coded categories of mobility. The conflation of boundaries as international political lines and as practices of identity construction signals the corporeal geographies of the nation-state. I have explored performances of distinction guided by Agamben's (1998) concepts of the operation of sovereign power that works paradoxical spaces at once inside and out, generated during times of crisis when the law is suspended, until this state of exception becomes the rule. Conditions of exclusion need not be as violent or as obvious as the suspension of the law or the isolation of the camp, but can be more insidious, pervasive, subtle, yet ultimately equally alienating. Repetitions of distinction in times of crisis give way to exception and exclusion. What will be the ramifications of performance of distinction for refugees and others on the move in search of protection and livelihood and their others – citizens – who are ostensibly provided with such protection and livelihood? If states continue to police national borders and corresponding categories of mobile identities and spaces of legal possibility, will the refugee or the refugee claimant disappear?

References

Agamben, G. (1998), *Homo Sacer: Sovereign Power and Bare Life* (Stanford: Stanford University Press).

Amoore, L. (2006), 'Biometric Borders: governing mobilities in the war on terror', *Political Geography* 25, 336–51.

Anderson, B. (1983), *Imagined Communities* (London and New York: Verso).

Anderson, K. (1991), *Vancouver's Chinatown: Racial Discourse in Canada, 1875–1980* (Montreal, Kingston: McGill-Queen's University Press).

Ang, I. (1999), 'Racial/Spatial Anxiety: "Asia" in the Psycho-Geography of Australian Whiteness', in G. Hage and R. Couch (eds.) *The Future of Australian Multiculturalism: reflections on the twentieth anniversary of Jean Martin's The Migrant Presence* (Sydney: Research Institute for Humanities and Social Sciences University of Sydney).

Appadurai, A. (2006), *Fear of Small Numbers* (Durham and London: Duke University Press).

Bassel, L. (2008) 'Citizenship as Interpellation: Refugee Women and the State,' *Government and Opposition* 43:2, 293-314.

Bhandar, D. (2008) 'Resistance, Detainment, Asylum: The Ontological Limits of Border Crossing in North America.' In D. Cowen and E. Gilbert (eds.), *War, Citizenship, Territory* (New York: Routledge), 281-302.

Butler, J. (1997), *Excitable Speech: A Politics of the Performative* (New York: Routledge).

Citizenship and Immigration Canada. (2001), 'Review of the Immigration Control Officer Network – Final Report.' http://www.cic.gc.ca/EnGLish/resources/audit/ico/index-e.asp (accessed 2 January 2010).

Clarkson, S. (2000), '600 is too many', *Ryerson Review of Journalism*, spring.

Cresswell, T. (1997), 'Weeds, plagues and bodily secretions: a geographical interpretation of metaphors of displacement', *Annals of the Association of American Geographers* 87:2, 330–45.

Cresswell, T. (2006), *On the Move: Mobility in the Modern Western World* (London and New York: Routledge).

Dauvergne, C. (2004), 'Sovereignty, Migration and the Rule of Law in Global Times', *Modern Law Review* 67:4, 588–615.

Ellis, M. and Wright, R. (1998), 'The Balkanization metaphor in the analysis of US immigration', *Annals of the Association of American Geographers* 88, 686-698.

Foucault, M. (1991), 'Governmentality'. In G. Burchell, C. Gordon, and P. Miller (eds.), *The Foucault Effect: Studies in Governmentality* (Chicago: University of Chicago Press), 87–104.

Foucault, M. (1995/1977*) Discipline and Punish: The Birth of the Prison* (New York: Vintage Books).

Greenberg, J. (2000), 'Opinion discourse and Canadian newspapers: the case of the Chinese boat people', *Canadian Journal of Communication* 25:4, 517–37.

Hage, G. (1998), *White Nation: fantasies of White supremacy in a multicultural society* (Annandale: Pluto Press).

Hier, S. and Greenberg, J. (2002), 'Constructing a discursive crisis: risk, problematization and *illegal* Chinese in Canada', *Ethnic and Racial Studies* 25(3), 490–513.

Honig, B. (2001), *Democracy and the Foreigner* (Princeton and Oxford: Princeton University Press).

Hyndman, Jennifer (2000), *Managing Displacement* (Minneapolis: University of Minnesota Press).

Hyndman, Jennifer (2004), 'The (Geo)Politics of Mobility and Access', in L. Staeheli, E. Kofman, and L. Peake, (eds.), *Mapping Women, Making Politics* (New York and London: Routledge) 169–84.

Kearney, M. (1991), 'Borders and Boundaries of State and Self at the End of Empire', *Journal of Historical Sociology* 4, 52–74.

Kristeva, J. (1982), *Powers of Horror: An Essay on Abjection* (New York: Columbia University Press).

Li, P. (1998), *The Chinese in Canada* (Oxford: Oxford University Press).

Mahtani, M. and Mountz, A. (2002), 'Immigration to British Columbia: Media Representation and Public Opinion,' Research on Immigration and Integration in the Metropolis, Working Paper No. 02-15. http://www.riim.metropolis.net/ (accessed 1 May 2003).

McGuinness, S. (2001), 'Canadian Print Media Coverage of the 1999 Fujian Migrants', Centre of Chinese Research, Institute of Asian Research, University of British Columbia.

Mountz, A. (2004), 'Embodying the nation-state: Canada's response to human smuggling', *Political Geography* 23, 323–45.

Mountz, A. (2010), *Seeking Asylum: Human Smuggling and Bureaucracy at the Border* (Minneapolis: University of Minnesota Press).

Nadig, A. (2002), 'Human Smuggling, National Security, and Refugee Protection', *Journal of Refugee Studies* 15, 1–25.

Nevins, J. (2002), *Operation Gatekeeper: The Rise of the "Illegal Alien" and the Making of the U.S.-Mexico Boundary* (New York and London: Routledge).

Ong, A. (2006), *Neoliberalism as exception: mutations in citizenship and sovereignty* (Durham: Duke University Press).

Perera, S. (2002), 'What is a Camp…?', *borderlands e-journal* 1:1, 1–15 (accessed 24 May 2004).

Pratt, G. in collaboration with the Philippine Women's Centre (1998), 'Inscribing Domestic Work on Filipina Bodies', in H. Nast and S. Pile (eds.) *Places through the Body* (London: Routledge), 283–304.

Province (1999), 'Enough Already. "It's time to toughen the law"', 1 September.

Province (1999), 'Quarantined,' 21 July.

Province (1999), 'Beware, illegal immigrants. We Canadians can be pretty ruthless', 13 August.

Razack, S. (1999), 'Making Canada White: Law and the Policing of Bodies of Colour in the 1990s', *Canadian Journal of Law & Society* 14:1, 159–84.

Rivers, Kevin and Associates (2000), 'Opinion Polls as Baseline Measures: A Discussion Paper', Prepared for Ministry of Multiculturalism and Immigration, Province of British Columbia.

Said, E. (1979), *Orientalism* (New York: Vintage Books).

Torpey, J. (2000), *The Invention of the Passport: Surveillance, Citizenship and the State* (Cambridge and New York: Cambridge University Press).

Vancouver Sun (1999), 'Detained aliens investigated,' 22 July.

Index